Creating STEM Lessons for Your Curriculum

Jo Anne Vasquez
Michael Comer
Joel Villegas

Heinemann
Portsmouth, NH

Heinemann
361 Hanover Street
Portsmouth, NH 03801–3912
www.heinemann.com

Offices and agents throughout the world

The authors and publisher wish to thank those who have generously given permission to reprint borrowed material:

Figure 1.1 and Appendix A: Whirligig Template from "Whirligig Lollapalooza: Exploring Science and Engineering Practices," Middle Grades Science Lesson by National Math + Science Initiative. Copyright © 2013 by National Math + Science Initiative. Reproduced with permission.

Figures 1.2, 1.4, and 4.4: From *STEM Lesson Essentials: Integrating Science, Technology, Engineering, and Mathematics* by Jo Anne Vasquez, Michael Comer, and Cary Sneider. Copyright © 2013 by Jo Anne Vasquez, Michael Comer, and Cary Sneider. Published by Heinemann, Portsmouth, NH. All rights reserved.

Cataloging-in-Publication Data is on file with the Library of Congress.

ISBN: 978-0-325-08776-4

Editor: Katherine Bryant
Production editor: Sonja S. Chapman
Typesetter: Cape Cod Compositors, Inc.
Cover and interior designs: Bernadette Skok
Manufacturing: Steve Bernier

Printed in the United States of America on acid-free paper
21 20 19 18 17 PAH 1 2 3 4 5

We dedicate this book to the tireless efforts of teachers everywhere
who persevere each day to create an atmosphere of restless curiosity
and inquisitiveness in their classrooms . . . fostering an environment
for learning and nurturing the imagination of their students.

We know what the journey is like and we are forever appreciative
and thankful for your dedication to improving the lives of those you touch.

Contents

Foreword

One of the best parts of my job is having the opportunity to visit many different schools across the country to see what science and STEM education look like in action. About a year ago, I visited a newly transitioned STEM elementary school. As I walked up the hallway, I heard lots of student excitement coming from several different directions. In the fourth-grade hallway, I saw groups of students working on a catapult that would launch a mini marshmallow. I asked the children what they were working on. They explained (all at once) the goal of the challenge and the constraints (their "rules") for making the contraption. They were using a variety of materials: tongue depressors, rubber bands, rulers, plastic spoons, tape, measuring tape, and so on. The children showed me the plan that they had drawn and then followed to build their contraption. They even had a budget for the cost of materials it took to make the device. They explained to me that this was not their first design, but a second version with improvements based upon their first trial. The most interesting conversation happened when the children had realized that they were losing some distance due to the arch of the pathway of the marshmallow. Clearly this was something that they were trying to figure out. As I was getting ready to move along to the next group, I couldn't help but notice one young man who was all smiles. I asked him why he was smiling so much. He said that he just loved school! I asked him why and he said that last year school was boring with just reading and writing. Now school involves these projects that they get to build and then figure out how to make them better. He said that there is still reading and writing, but it all relates to the projects that they are working on—"it is just more fun," he said.

Although STEM has been around for quite some time, in the last decade the movement has gained traction. Funding and policy agencies and national trends have championed STEM education to help address issues related to societal and global concerns, economic recovery, and workforce development. At the same time, others are taking advantage of the momentum of STEM to promote even more integration between subjects, giving rise to acronyms like STEAM (Science, Technology, Engineering, Arts, and Mathematics), STREAM (Science Technology, Reading, Engineering, Arts, and Mathematics), or HAMSTER (Humanities, Arts, Mathematics, Science, Technology, Engineering, and Research). In these trends we see the value of integrating traditionally siloed disciplines as a way of making learning more authentic for students. One needs to look no further than some of the latest research and newest technologies to see that the traditional silos of education are antiquated. Research in prosthetics relies on physics, biology, technology, engineering, and mathematics to build devices to assist amputees. In the medical field, chemistry, biology, physics, engineering, and mathematics come together to develop new medicines with greater efficacy, fewer side effects, and new delivery systems. Science learning is not done in isolation, but should be considered as a symbiotic and synergistic dynamic, looking at real world problems and collecting evidence to build and communicate claims and solutions that are truly interdisciplinary endeavors. STEM instruction must begin building those integrated skills of problem solving and critical thinking in pre-kindergarten and continue that instruction in a concerted effort through high school in order to have a science-minded citizenry that is prepared to help solve global issues.

In *STEM Lesson Essentials*, two of the authors of this book, along with Cary Sneider, provided an invaluable primer for STEM Education as a professional development tool. Both practicing and pre-service teachers have used the book to begin to understand what STEM teaching is, the reasoning behind it, and the practices associated with it, and were introduced to a few ways to begin to integrate the STEM disciplines in their classroom instruction. In this second book, *STEM Lesson Guideposts: Creating STEM Lessons for Your Curriculum*, Jo Anne Vasquez, Michael Comer, and Joel Villegas continue the journey to understanding STEM education and how it can be implemented in classroom settings, moving beyond the acronym to educational practice. This book offers a strategy for building integrated STEM lessons and units of instruction through the W.H.E.R.E. framework. Readers will begin their journey of STEM lesson and unit design by identifying *what* students will learns and *why*, which reinforces the practices and guiding principles discussed in *STEM Lesson Essentials.* The second leg of the journey is the *how*, guiding readers to consider the different experiences that are needed for students to engage in STEM learning and determining what *evidence* and *evaluation* is required to make sure that student learning is progressing in the right direction. The third leg of the journey includes *rigor* and *relevance*, connecting their students' background and experiences to the learning. The journey continues with *excitement*, *engagement*, and *exploration*, ensuring that students can make the deeper connections that are possible through integrated STEM learning. Finally, the last two chapters provide examples of schools where STEM education is working and what those schools had to go through to become successful.

This book provides specific examples as models for using the W.H.E.R.E. framework. Its organized format can help you create integrated lessons and units so that your students are actively engaged in learning and developing the skills necessary to be successful for their future. Once again, the authors have successfully moved STEM from an idea to an actionable model of real educational practice.

—David T. Crowther
Professor Science Education
Executive Director: Raggio Research Center for STEM Education
University of Nevada, Reno

Acknowledgments

A project like this is never a solo task and we are sincerely appreciative of all the extraordinary educators who helped support us, added their wise counsel, and made suggestions as we developed the W.H.E.R.E. model. Most of all, thank you for cheering us on through the development and refinement of *STEM Lesson Guideposts*.

Thanks to our team at Heinemann: Katherine Bryant, our editor, who was always there and provided thoughtful suggestions; Sonja Chapman, Production Editor, who helped to "tweak" the design of the pages; and to Josh Evans, Product Manager, for his work in positioning our book in the marketplace. Also, thank you to Lorraine Smith-Phelan who copyedited our initial manuscript.

There have been many teachers and instructional leaders around the country who participated in the professional learning experiences where we field-tested and revised the Guidepost template, refined the strategies, and improved the W.H.E.R.E. Model. These educators shared their collective wisdom through frank feedback and critical evaluation of our work. We would like to acknowledge Darrel McDaniel, STEM Education Development Leader, Calcasieu Parish Schools, Lake Charles, Louisiana. Darrel and his STEM Master Teachers and coaches provided us a testbed in which to incubate our strategies from the very beginning. They gave us constructive feedback and provided many creative ideas for the STEM lessons.

Special thanks to the teachers and leaders of W. F. Killip Elementary in Flagstaff, Arizona. The story of their STEM learning journey is shared in Chapter 8. Principal, Joe Gutierrez, and lead STEM Coordinator, Ted Komada, were instrumental in guiding their teachers through the development of the Rover unit using the STEM Lesson Guideposts.

Chapter 9 highlights the strategies used to devise a systemic approach to implementing a STEM initiative as told by Dr. Larry Plank, Director of STEM Education for Hillsborough County, Florida. Thank you, Larry, for your guidance, contribution, and wisdom during the development of this book. Your leadership and words remind us that any new and innovative educational strategy will soon fall by the wayside unless all participants have a common understanding and buy-in to the goals of what you are trying to accomplish.

We have also taken these strategies to Thailand as STEM professional development providers where we were part of the country's STEM Education Initiative. We used *STEM Lesson Essentials* and *STEM Lesson Guideposts* as the basis for training their STEM coaches and teacher leaders. Special thanks and appreciation to Dr. Pornpun Waitayangkoon, President of the Institute for the Promotion of Science and Technology (IPST) and to her skilled staff for their suggestions and guidance in the development of the Guidepost strategies.

Finally, thanks to Dr. David Crowther, Executive Director Raggio Research Center for STEM Education, University of Nevada, Reno. David has been a champion of STEM and Science Education. He agreed to write the foreword for this book even with his hectic schedule of research, teaching, and being the 2017 President of the National Science Teachers Association. Thank you so much.

Introduction and Overview

The STEM Lesson Guideposts Learning Journey

Today's economy demands that all young people develop high level literacy, quantitative reasoning, problem solving, communication and collaboration skills, all grounded in a rigorous, content-rich K–12 Curriculum. Acquiring this knowledge and these skills ensures that high school graduates are academically prepared to pursue the future of their choosing.

Closing the Expectations Gap: 2014 Annual Report on the Alignment of State K–12 Policies and Practice with the Demands of College and Careers
(Achieve 2015, 2)

Everywhere we turn, STEM (science, technology, engineering, and mathematics) has become the "soup du jour" in education. Curriculum leaders and policy makers are racing to implement STEM education. There are also those who think this is just a fad, a flash in the pan, and it too will go away. Today's increasingly technological world, with access to immense pools of knowledge, requires a citizenry that can discern real facts, recognize bias, and make decisions on a wide variety of issues. Understanding issues like the causes of climate change, the impact of industrial development in an environmentally sensitive area, or the pros and cons of alternative energy production are required of an informed and educated electorate. We also want to equip students with the skills and knowledge needed to be successful in careers of their choosing. A workforce prepared for the twenty-first century must be proficient in skills that include oral and written communication, critical thinking and problem solving, professionalism and work ethic, teamwork and collaboration, working in diverse teams, applying technology, and leadership and project management. In other words, they need to have an educational foundation that is built upon real-world, rigorous, and relevant learning experiences: exactly what integrated STEM teaching provides.

With a well-developed STEM learning experience, students won't have to ask, "Why do I have to learn this?" because the STEM experience itself provides them with the answer. Such experiences engage students with multiple opportunities to apply the skills and knowledge they have learned in creative, rigorous, and most importantly, relevant ways.

STEM: From What to How

In the 1990s, the acronym *STEM* was introduced by the National Science Foundation as it continued to receive interdisciplinary proposals that overlapped the directorates within the foundation. In 2007, the National Science Board, the governing board of the National Science Foundation, released *A National Action Plan for Addressing the Critical Needs of the U.S. Science, Technology, Engineering, and Mathematics Education System*, putting STEM teaching and learning directly in the forefront of our educational system. Government and private funding began to flow toward all types of STEM education programs, and STEM became recognized as a meta-discipline—an integration of formerly separate subjects into a new and coherent field of study.

But everywhere you turned, schools, teachers, and districts had questions:

- "Why add engineering to an already overloaded schedule?"
- "Where do I find time to teach something new?"
- "How do I integrate the different disciplines?"
- "What is considered technology?"
- "Where can we fit this into our busy reading and math schedules?"
- "How can we evaluate what we are doing and give our students a grade?"

These questions reflected an underlying confusion over what teaching STEM really meant. To address this confusion, two of us and our colleague Cary Sneider wrote *STEM Lesson Essentials* in 2013. In it, we defined STEM this way:

> *STEM is an interdisciplinary approach to learning which removes the traditional barriers separating the four disciplines of science, technology, engineering, and*

mathematics, and integrates them into real world, rigorous and relevant learning experiences for students. (Vasquez, Sneider, and Comer 2013)

This book builds on that definition, the *what* of STEM, to show you *how* you can make this interdisciplinary approach happen in your classroom, school, or district. We will help you answer questions like these: How can we create our own STEM units? How do we address the standards we need to cover? How can we create experiences that will engage all students? What does rigor look like in a lesson? And, most importantly, how do we begin? Understanding what STEM education is was only the first step. Knowing how is next.

In this book, we introduce to you a model and a template for planning STEM units. The W.H.E.R.E. Model includes five key components, or as we call them, guideposts:

- *W—What* and *why:* What needs to be learned and why?
- *H—How:* How do I plan to get there? What experiences do students need to have?
- *E—Evidence* and *evaluation:* What evidence of learning will be used and how will I evaluate the final product or project?
- *R—Rigor* and *relevance:* How will I provide opportunities to increase the rigor of students' thinking and the relevance to students' experiences in the real world?
- *E—Excite*, *engage*, and *explore:* How do I excite the learners, cognitively engage them, and allow for them to explore for deeper understanding of the content and skills?

Overview of the STEM Lesson Guideposts Learning Journey

Chapter 1: Beginning the Journey

This chapter reviews the levels of STEM integration through the lens of a Whirligig activity and looks back at the STEM practices detailed in *STEM Lesson Essentials.*

Chapter 2: Planning the STEM Journey

This chapter introduces the W.H.E.R.E. Model and the STEM Lesson Guidepost Planning Template with an overview of its structure.

Chapter 3: The *W*—Picking the Destination

This chapter defines the W Guidepost: *What* needs to be learned and *why* the students would be interested in learning this content and skills. It models the process through the *Engineering a Hat* STEM unit and highlights the importance of starting with the standards, knowledge, skills, and key understandings that the unit will address.

Chapter 4: The *H* and the *E*—Planning and Evaluating the Route

This chapter describes the H and E Guideposts: *How* are we going to get there? What *evidence* of learning will we use to measure success? The process of planning the sequence of instruction and determining evidence of success is modeled through the *Wind Turbine* STEM unit.

Chapter 5: The *R* and *E*—Keeping the Trip Interesting

This chapter develops the R and E Guideposts: improving the *rigor* and increasing the *relevance* in the learning journey and addressing the question of how we will *excite* and *engage* the students and facilitate their inquiry *explorations.* All of this is modeled through the *Feeding Fluffy* STEM unit.

Chapter 6: Changing Drivers

This chapter models how changing the primary content focus (the "driver") impacts the direction of the unit. We revisit the *Wind Turbine* STEM unit and see how the change of drivers affects the STEM Lesson Guidepost Planning Template and alters the STEM unit experience.

Chapter 7: Now It's Your Turn

This chapter puts you in the driver's seat so you can practice modifying one of the STEM units previously introduced. It guides you through the process and ends with you creating your own template for a new STEM unit called the *Art Pedestal Project.*

Chapter 8: Changing the System

This chapter introduces the Concerns-Based Adoption Model (CBAM), a model of change, and explores the development of STEM units in a school over the course of three years. It describes the process as the school moved from STEM awareness to coaching and mentoring of teachers and concludes with the journey of the third-grade team as it used the guideposts to create and implement its own STEM unit.

Chapter 9: Creating a Pathway

In this chapter, we will look at Hillsborough, Florida, a large county public school district, whose approach to creating a systemic STEM initiative highlights the need to have all stakeholders involved in the process of developing and implementing a successful STEM program.

At the end of every chapter you will find a "Think About It" section. This concluding reflection paragraph provides an overview of the chapter and gives you an opportunity to think about your own classroom experiences. This section can also be a focus for discussions in professional learning communities to work collaboratively and improve your own knowledge and skills. To further guide those discussions, we've added a short series of questions we've called "Speed Bumps." Based on our experiences in leading professional growth workshops, these questions represent some of the challenges presented by teachers as they begin to move ahead to implement changes in their curriculum. The questions are designed to generate thoughtful conversations and refer back to the specific sections found in the chapter. Both the "Think About It" and the "Speed Bumps" are meant to provide a time to hit the pause button and have a dialogue with your team or study group. We hope you enjoy this feature of our book.

Think About It

Integrated and interdisciplinary teaching has been around for many years, but the focus now on STEM learning has awakened a whole new generation of teachers to the possibility of exciting their students by helping them make connections between the disciplines they are studying and applying this new learning in new and novel ways. STEM is about the application of learning. Join us as we travel through this STEM learning journey, where we will champion your professional growth as you learn how to develop your own STEM units.

Speed Bumps

- How acquainted are you with STEM teaching?
- Are you implementing STEM units into your classroom or school now? What has made it successful? What more do you hope to learn?
- What types of professional development sessions have you attended? How have they provided a foundation for your understanding of STEM teaching and learning? What do you feel still needs more attention?
- Have you developed your own STEM units? What obstacles, if any, have you had to overcome? What support would make that process more efficient or rewarding?

Beginning the Journey

Moving Beyond the Acronym

As hard as it is, defining STEM is the easy part; implementing STEM education on a large scale is the hard part.

—Rodger Bybee, Executive Director (Retired) Biological Science Curriculum Study, personal communication, March 2012.

STEM teaching and learning gives all students opportunities to apply the skills and knowledge they are learning in new and different ways, thus providing the answer to the age-old question. "Why do I need to learn this and when will I ever use it?" But what does STEM integration really mean?

Understanding Integration

Integration literally means to combine separate parts into a whole. Thus, learning in an integrated way shifts the emphasis to the comprehensive, connected understanding of whole concepts. Research on cognition and learning shows that the organization of knowledge, the ability to make connections between concepts and representations, is key to the development of expertise in a domain. Experts not only know more about a domain, but also understand how ideas and concepts are related to each other and their relative importance and usefulness. Experts are also able to recognize and organize patterns of information and use them to solve problems. Experts can attend to the structural aspects of a new problem and thus connect the new tasks or concepts to prior experiences more readily and more meaningfully.

But what does integration mean for STEM teaching and learning? Is it sufficient for students to just recognize connections between concepts in different fields? Or does integration imply the assimilation of concepts from two or more different fields so they become one? Should integration involve skills as well as concepts? How about connections to other domains (social studies, language arts, the performing arts, and others)? Should integration include aspects of everyday life and real-world contexts? Depending on the circumstances, all of these ideas may be part of defining integration. Often the content may begin with the skills and knowledge from one discipline that then intersect with the skills and knowledge of another, such that an integrated approach evolves where the lines between the disciplines become blurred. Research supports the position that it is important for integrated STEM experiences to help students build knowledge in the individual disciplines and equally important to help students make connections among these disciplines (National Research Council 2014b). This connection building helps students take the knowledge and skills learned in one context and apply them in another.

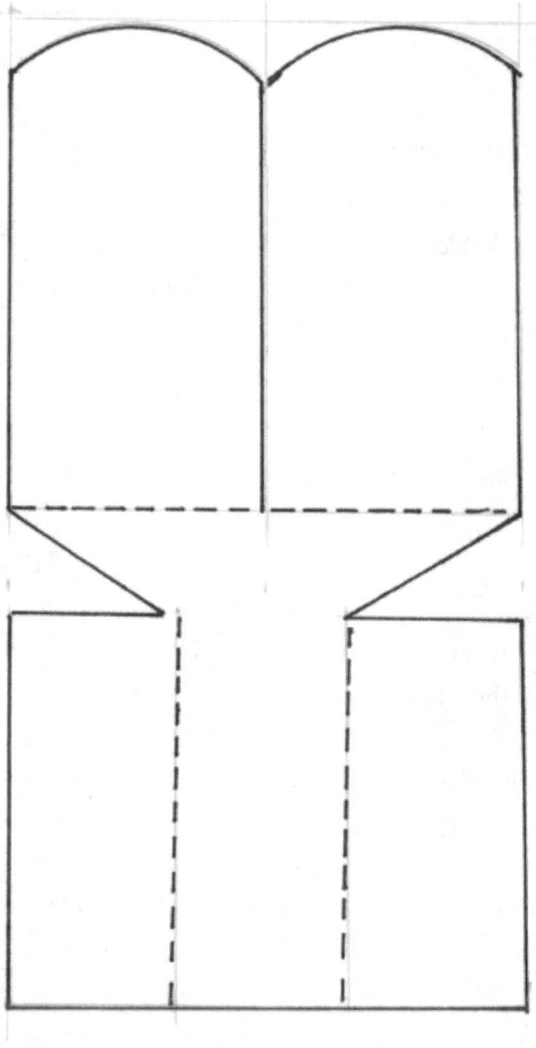

FIGURE 1.1 Whirligig Pattern

Throughout the discussion that follows, we will use examples from a STEM unit called *Whirligigs.* This unit is based on the National Math + Science Initiative unit called *Whirligig Lollapalooza*. The whirligig (Figure 1.1) is a paper device with two wings and is crafted in a way to float in a spiral as it descends. Students explore engineering design concepts as they modify the device and evaluate how their changes impact the dynamics of the whirligig's descent. A copy of the Whirligig Pattern is in Appendix A.

Levels of Integration

You may think of integrated STEM teaching as problem-based or project-based learning. This level of integrated experience is often the most difficult to achieve, because it takes careful planning, collaboration, and time to execute within the classroom setting. It is not, however, the only useful way to approach integration. We have developed three distinct levels of or approaches to integration: *multidisciplinary* or *thematic integration*, *interdisciplinary integration*, and *transdisciplinary integration*. We'll discuss these levels briefly here; for more detail, see *STEM Lesson Essentials* (Vasquez, Sneider, and Comer 2013).

The image of an inclined plane in Figure 1.2 will help you visualize the different and increasing levels of STEM integration.

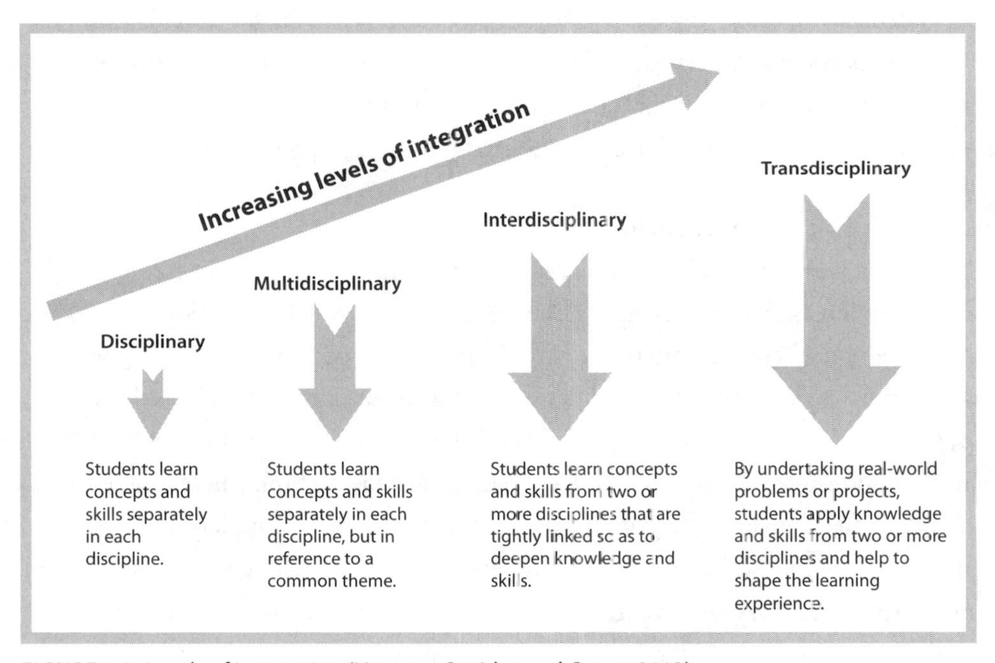

FIGURE 1.2 Levels of Integration (Vasquez, Sneider, and Comer 2013)

Disciplinary Teaching

At the beginning of this plane sits *disciplinary* teaching, in which the students learn the content and skills of individual subject areas separately. Here the presentation and assimilation of subject-specific content is the ultimate goal. Accumulation of information often drives the learning experience without much emphasis on how or where to use that information.

In the case of the *Whirligigs* unit, the understanding in this disciplinary approach might focus only on the science content. Students could investigate several different parameters that affect the whirligig flight, but the focus would be solely on understanding the specific science concepts involving forces and motion: Does the height of the drop make a difference? Does the whirligig always spin in the same direction? Why?

The disciplinary approach may include aspects of other disciplines in the activities the students perform, but the *learning objectives* are solely in support of the single discipline. As the students are asked to explore the various factors that impact the flight of the whirligig, they focus solely on the development of the investigation and the acquisition of the science understanding and not necessarily on the aspects of measurement or design changes. The learning is just about the science. Although the questions might move students toward the engineering/design process, it is not the focus. This kind of teaching can become a foundation for more integrated STEM experiences, however; through relevant STEM experiences, students can apply this knowledge in new ways that truly demonstrate that they have learned the content and can apply those skills.

Multidisciplinary Integration

The next level is *multidisciplinary* integration, in which a unifying theme connects different content areas. The acquisition of content-specific knowledge and skills is still the learning goal, but there is an attempt to loosely link the learning together using broad-based attributes or general commonalities. Multidisciplinary integration has advantages, particularly when a team of teachers is initially planning a STEM unit where each teacher is responsible for his or her own content area of instruction. There is an explicit intent within the instruction to connect the knowledge, skills, and understandings together in a way that will engage and make sense to students. These types of integrated learning experiences may also eventually lead to encouraging an increased level of integration as a natural outgrowth of initial shared planning.

In the whirligig example, the multidisciplinary approach might connect the individual disciplines by organizing the curriculum around a common theme like "Aviation: Its Impact on Civilization." As shown in Figure 1.3, instruction is centered on supporting the individual content curriculum guidelines for each discipline, and the theme is largely tangential. In mathematics, for example, students might happen to use aviation data to address the data analysis and statistical variability standards. In English, student readings might focus on nonfiction text about the lives of key aviation pioneers. In science, the big idea of force and interactions is examined through experimentation on the forces that affect the flight of the whirligig, and the Career and Technical Education (CTE) or Ed-Tech course might focus on engineering design principles of aircraft. In health, the understandings of body systems might examine the effects of long distance flying on circulation, and in social studies, students studying World War II might examine how the use of aircraft influenced key events and possibly the outcome of the war. In each case, the discipline uses the theme to advance the preexisting learning objectives of the curriculum, but not necessarily fusing the learning objectives across the domains.

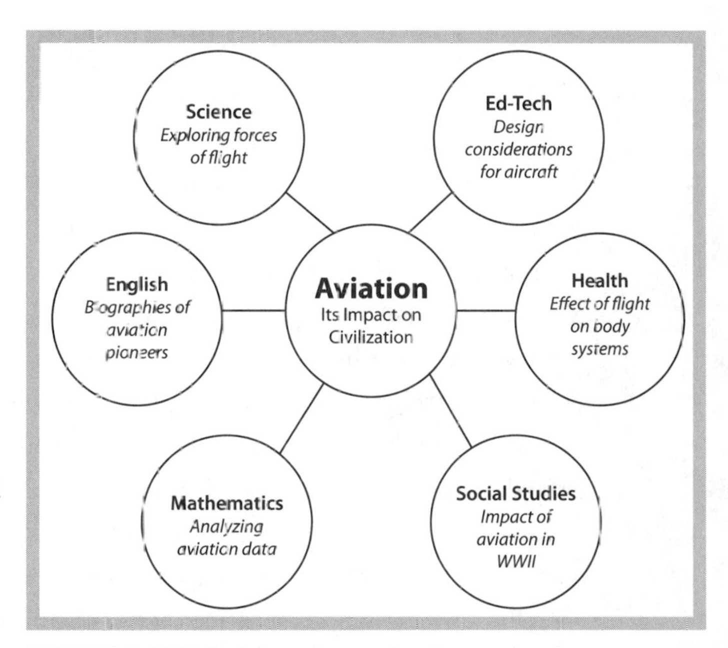

FIGURE 1.3 Multidisciplinary Integration Approach

Interdisciplinary Integration

In *interdisciplinary* STEM integration, the connections between the content and skills become much more explicit and the learning more interdependent. In science, for example, the students may be investigating the forces on and motion of an object rolling down a ramp they have constructed. Although they intentionally apply their math skills to record time and distance, analyze and interpret data, and develop charts and graphs to represent their findings, the focus of learning is on the "big idea." The students are applying understandings from each content area without separating the learning into the individual discipline silos, and they are concentrating on the broader conceptual connection between the two disciplines. The tasks are integrated, interdisciplinary, relevant, and rigorous.

In the interdisciplinary approach to the *Whirligigs* unit, the major goal of the learning focuses on the big idea of *cause and effect*. Throughout the whirligig investigation, the students experience how changing certain variables (such as wing length, number of paper clips, or materials used) will have an effect on the performance of the whirligig. In mathematics, through the experimental data of their investigations, students will come to understand how to solve equations involving variables and to quantitatively analyze the relationship between the independent and dependent variables. They continually ask the question, "How do changes to the variables affect the data that we generate?" Students may apply engineering design principles to the tasks of modifying the whirligig to achieve a specific desired result.

The learning is focused around the shared common understandings that are embedded in the curriculum for both mathematics and science. However, that understanding can be expanded to other disciplines as well. The same common learning, "cause and effect," can be linked to English language arts and social studies. Students begin to experience the natural connection of these crosscutting concepts, understanding how the learning is transferred across disciplines through natural connections, regardless of the topic being studied.

Transdisciplinary Integration

Within the *transdisciplinary* approach, teachers bring together real-world application and problem-solving strategies. The learning focuses around Essential Questions, led by one "Driving Question" that provides the focus. Typically, transdisciplinary units start with a scenario that provides the rationale or the challenge for students to pursue. This is often a design or engineering challenge, but it does not have to be; it may focus on answering a question instead.

In the case of the *Whirligigs* unit, we might start with this scenario and Driving Question:

> *A team of scientists and engineers is planning to land a robot spacecraft on the surface of Mars. They already have the Curiosity and Spirit surface rovers working for them but have decided to land another robot spacecraft in a different location. The conditions in this new location are very treacherous and the surface is uneven with only one place where the spacecraft can land. If the spacecraft is traveling too quickly when it lands, it will crash, and if it goes too slowly, it will miss the landing spot.*

THE DRIVING QUESTION

As engineers, how do we design and build a whirligig spacecraft that can descend from a height of three meters and land in an exact spot within a time window of three seconds?

Now the students will embark on the same type of learning experiences that were described in the earlier integration levels. The difference is that the emphasis at the transdisciplinary level is not only the acquisition of content-specific knowledge or the application of discrete discipline skills, but also blending together objectives that focus on applying the key understandings, using multiple skills and different types of content knowledge in a new and novel way.

Recognizing that there are distinct levels of integration is important in helping you and your team to begin planning your own STEM units. Remember, you don't always have to jump right into the transdisciplinary level. You can certainly start with just beginning to integrate two or three subjects around a common skill or concept (interdisciplinary). Or you might even begin with a multidisciplinary approach built around a common theme connecting the different subject areas.

Practices: The Ties That Bind STEM Learning Together

Throughout an integrated STEM unit, we want students to be engaged in the process of doing the science, math, and engineering while using and applying technology and its principles. This process reflects a series of practices or skills that we want students to use as they explore their world and that engage students cognitively in their learning. Many of these practices involve covert processes that are hard to monitor. Luckily, each of the STEM disciplines (and others, such as English language arts) has a set of practices that proficient thinkers exhibit as they engage cognitively in the content.

These practices have been codified in various standards, including the Next Generation Science Standards and the Common Core State Standards for Mathematics. Comparing them as in Figure 1.4 shows how the practices from different disciplines are consistent with and parallel to one another.

Science	Engineering	Technology	Mathematics
Ask questions.	Define problems.	Become aware of the web of technological systems on which society depends.	Make sense of problems and persevere in solving them.
Develop and use models.	Develop and use models.		Model with mathematics.
Plan and carry out investigations.	Plan and carry out investigations.	Learn how to use new technologies as they become available.	Use appropriate tools strategically.
Analyze and interpret data.	Analyze and interpret data.		Attend to precision.
Use mathematics and computational thinking.	Use mathematics and computational thinking.	Recognize the role that technology plays in the advancement of science and engineering.	Reason abstractly and quantitatively.
Construct explanations.	Design solutions.		Look for and make use of structure.
Engage in argument from evidence.	Engage in argument from evidence.	Make informed decisions about technology, given its relationship to society and the environment.	Construct viable arguments and critique the reasoning of others.
Obtain, evaluate, and communicate information.	Obtain, evaluate, and communicate information.		Look for and express regularity in repeated reasoning.

FIGURE 1.4 Practices Chart, *STEM Lesson Essentials* (Vasquez, Sneider, and Comer 2013, 38)

Think About It

As you have seen in this chapter, although each level of integration has its own unique position in the learning continuum, we believe that creating an integrated transdisciplinary STEM unit provides a far richer and more meaningful learning experience for the student. In the chapters to come, we will provide tools to help you plan these kinds of experiences.

Speed Bumps

- How would you summarize the differences between the levels of integration?
- What advantages does each level of integration offer?
- What are some of the challenges to implementing each level?
- Describe a lesson or unit you teach and tell what level of integration it represents. How might it look if it were developed for a different level of integration?
- As you have explored the different levels of STEM integration, where do you see you and your team starting your STEM unit planning?

Planning the STEM Journey

W.H.E.R.E. Are Your STEM Guideposts?

At its core, teaching is an art that calls on its practitioners to work simultaneously in multiple media, with multiple elements. Central to our STEM teaching journey is what we ought to teach and what we want our students to know, understand, and be able to do. Coherent, meaning-rich curriculum provides opportunities for the learning process to happen within students, not to them.

—Vasquez, Sneider, and Comer, *STEM Lesson Essentials*

Over the last couple of years, many of the educators we have been fortunate to work with have developed an understanding of STEM teaching and learning. Now they are beginning to ask, "How can we develop our own STEM lessons and units? How do we address the standards we need to cover? How can we create experiences that will engage all students? What does rigor look like in a lesson? And how do we begin to make these experiences relevant to all students?" Understanding *what* STEM education is was only the first step; understanding *how* to craft STEM experiences is the second step in moving toward implementing STEM into the classroom. In this chapter, we introduce a model that can help you take that next step.

W.H.E.R.E. Do We Begin?

We have heard so many teachers ask, "Where do I begin?" Inspired by that question, we devised the W.H.E.R.E. Model as a guide through the process for creating a cohesive and integrated STEM unit. This model provides a simple mnemonic that identifies the five essential building blocks of unit development:

- *W*—what and why. What needs to be learned and why?
- *H*—how. How do I plan to meet this goal?
- *E*—evidence and evaluate. What evidence of learning will be used and how will I evaluate the final product?
- *R*—rigor and relevance. How will I provide opportunities that will lead to increased rigor and relevance?
- *E*—excite, engage, and explore. What exploration activities will excite the students and engage them in the STEM practices?

We refer to each letter of the W.H.E.R.E. Model as a "guidepost." Like highway marker signs, the guideposts provide direction and key information at critical times when planning your STEM journey. Each guidepost represents a decision point in the planning process, an opportunity to reflect on the critical ingredients that make for effective instruction.

Big Ideas and Backward Design

The W.H.E.R.E. Model is based on current research and best practices for curriculum development that leads to deep student understanding. As described by Jay McTighe and Grant Wiggins in their *Understanding by Design Framework* (2013), understanding is revealed when students make sense of and transfer their learning through authentic performance tasks. The authors suggest six facets that serve as indicators of understanding:

- the capacity to explain
- the ability to interpret
- understanding how to apply new knowledge and skills
- the ability to shift perspective
- the ability to empathize
- the ability to self-assess.

An effective STEM unit embodies these six facets as essential goals of its instruction. The STEM Lesson Guidepost Template captures these ideas within the individual guideposts that lead to effective planning. To this end, we have adapted and expanded the three-stage backward design process developed by McTighe and Wiggins (2005).

The planning highlights three key stages:

1. Identify desired results:
   - What should students know, understand, and be able to do?
   - What content and skills will be necessary for student understanding?
   - What are the big ideas, key concepts, knowledge, and skills that describe what the students will know and be able to do?
2. Determine acceptable evidence for assessment:
   - How will you know whether your students have achieved the desired results?
   - What will serve as evidence of student understanding and proficiency?
3. Plan the learning experiences:
   - What prior knowledge and skills will the students need to perform successfully?
   - What activities foster integration?
   - What levels of integration will be most effective to accomplish the learning goals?
   - How will there be opportunities for all students to participate?
   - What resources and materials will be needed to accomplish the goals?

We have expanded these three stages into the five guideposts found in the W.H.E.R.E. Model.

The W.H.E.R.E. Model

The W.H.E.R.E. Model in the STEM Lesson Guidepost Planning Template (Figure 2.1) provides a framework for planning STEM units by identifying strategic guiding questions and helping you navigate the five key areas—your guideposts—for planning an effective STEM unit. Each guidepost is introduced with a question that frames the purpose of that section and provides direction for the information that should be developed there.

STEM Lesson Guideposts Planning Template

WHAT are the desired results, including big ideas, content standards, knowledge, and skills?

List the content standards and what the students will know and be able to do.

WHY would the students care about this knowledge and these skills?

Craft the Driving Question that will lead to the development of the integrated tasks that provide for the application of the content, knowledge, and skills.

List the Essential Questions that can be answered as a result of the learning.

HOW do I plan to meet this goal?

Identify the pathway, including major tasks and milestones that result in answering the Driving Question.

E

EVIDENCE and EVALUATE: What evidence of learning will be used and how will I evaluate the final product?

PRE-ASSESSMENT: What prior knowledge is needed for this task?

Identify the prerequisite skills and understandings.

FORMATIVE: How will I measure student progress toward understanding?

Establish the assessment tools you will use to monitor progress and inform instruction.

SUMMATIVE: What criteria are needed for students to demonstrate understanding of the standards, content, and skills?

Create a checklist of criteria for use in a rubric.

RIGOR: How can I increase students' cognitive thinking?

Identify tasks that can elevate student thinking, improve inquiry, and increase conceptual understanding.

RELEVANCE: Does the learning experience provide for relevant and real-world experiences?

Identify current topics and local issues that can make the tasks more engaging.

EXCITE: What is the hook to excite the learner?

Create the scenario to engage the learner.

ENGAGE: How will the students be cognitively engaged throughout the unit?

List the STEM practices that will be used as evidence.

EXPLORE: What activities will help students address the Driving Question?

List questions for students to investigate that will lead them to a deeper understanding of the content and skills.

FIGURE 2.1 STEM Lesson Guideposts Planning Template. The complete planning template is found in Appendix B.

The following provides an overview of the individual guideposts and their strategic guiding questions. These guiding questions can be used as prompts for your thinking and brainstorming as you begin to develop your own STEM units, and we will expand on their ideas in the chapters to come.

W = *What* and *Why:* What Needs to Be Learned and Why?

The W Guidepost represents the *what* and the *why* in planning the instruction. (See Figure 2.2.) This section reflects stage 1 in the Understanding by Design (UbD) model mentioned earlier, focusing on the desired results of the learning experience—the *what*, and relatedly, the *why*. This guidepost helps to focus the instructional planning on the critical elements of applying content knowledge and skills to the real world. It hones the crafting of the Driving Question, which should capture and communicate the purpose of the learning. We will discuss the development of this guidepost further in Chapter 3.

Strategic Guiding Questions

- What are the desired results, including the "big ideas," content standards, knowledge, and skills (e.g., twenty-first-century skills) that this lesson/unit will teach?
- Why would the students need to know and understand these concepts and skills?
- What are the Essential Questions that will lead to student understanding of the content?
- What is the Driving Question that will lead to the development of the integrated tasks that provide the application of the content, knowledge, and skills?

WHAT are the desired results, including big ideas, content standards, knowledge, and skills?

List the content standards and what the students will know and be able to do.

WHY would the students care about this knowledge and these skills?

Craft the Driving Question that will lead to the development of the integrated tasks that provide for the application of the content, knowledge, and skills.

List the Essential Questions that can be answered as a result of the learning.

FIGURE 2.2

H = *How:* How Do I Plan to Meet This Goal?

The H Guidepost represents developing the pathway for designing STEM experiences. (See Figure 2.3.) *How* do I plan to meet this goal? What are the core experiences that the students will engage in to demonstrate understanding of the big idea? This guidepost helps you identify planning needs, calling out the high-level learning milestones that students should meet as they progress through the unit, and the sequence of learning experiences that will scaffold their learning to meet those milestones. The H Guidepost also provides an opportunity to begin thinking about the prerequisite knowledge or skills that students would need for each of these learning milestones. We will discuss the development of this guidepost further in Chapter 4.

Strategic Guiding Questions

- What learning experiences will enable the students to understand the concepts and skills in meaningful ways?
- How will the sequence of these learning experiences help students to construct their understanding and apply the skills?
- What are the major tasks or milestones that will lead toward answering the Driving Question?

HOW do I plan to meet this goal?

Identify the pathway, including major tasks and milestones that result in answering the Driving Question.

FIGURE 2.3

E = Evidence and *Evaluation:* What Evidence of Learning Will Be Used and How Will I Evaluate the Final Product?

The first E Guidepost represents planning for evaluating student understanding throughout the unit and is related to the preceding H Guidepost. (See Figure 2.4.) It focuses on what evidence will demonstrate that the student understands the concepts, knowledge, and skills as determined by the W Guidepost. Planning the assessment includes not only the final summative assessment, but also what formative assessments and preassessments should be used to monitor student progress toward the goals of the unit. We will discuss this further in Chapter 4 along with the H Guidepost.

Strategic Guiding Questions

- What prior knowledge and skills will the students need to demonstrate?
- What formative assessments will be used to measure student progress toward understanding and to inform instruction?

- What summative assessment, culminating product, or task(s) will demonstrate students' understanding of the standards, content, and skills?
- What criteria and tool(s) will be used for assessing student success?

E

EVIDENCE and **EVALUATE:** What evidence of learning will be used and how will I evaluate the final product?

PRE-ASSESSMENT: What prior knowledge is needed for this task?

Identify the prerequisite skills and understandings.

FORMATIVE: How will I measure student progress toward understanding?

Establish the assessment tools you will use to monitor progress and inform instruction.

SUMMATIVE: What criteria are needed for students to demonstrate understanding of the standards, content, and skills?

Create a checklist of criteria for use in a rubric.

FIGURE 2.4

R = *Rigor* and *Relevance:* How Will I Provide Opportunities That Will Lead to Increased Rigor and Relevance?

The R Guidepost reflects the research of Willard Daggett and draws on the Rigor and Relevance Framework developed by the International Center for Leadership in Education. Focusing on rigor and relevance moves the instruction from teacher-led to student-driven and shifts the perspective from a classroom context to a real-world setting. The R Guidepost emphasizes the different factors that can increase a student's cognitive involvement while at the same time making the learning experience more meaningful. (See Figure 2.5.) We will discuss the Rigor and Relevance Framework and how it relates to STEM learning as we explore the R Guidepost in Chapter 5.

Strategic Guiding Questions

- How will the learning experiences extend the student's thinking?
- What intriguing questions will foster greater inquiry to increase conceptual understanding?
- What opportunities will provide increased relevance and real-world connections for students?

RIGOR: How can I increase students' cognitive thinking?

Identify tasks that can elevate student thinking, improve inquiry, and increase conceptual understanding.

RELEVANCE: Does the learning experience provide for relevant and real-world experiences?

Identify current topics and local issues that can make the tasks more engaging.

FIGURE 2.5

E = Excite, Engage, and *Explore:* What Exploration Activities Will Excite the Students and Engage Them in the STEM Practices?

The second E Guidepost focuses on ways to engage and excite students as they navigate the unit. (See Figure 2.6.) In the UbD model, stage 3 identifies the importance of planning these learning experiences with emphasis on engaging student interest and addressing the needs of all. In the W.H.E.R.E. Model, this section also focuses on moving beyond the physical and social engagement of the classroom experience and highlights the importance of cognitive engagement.

Cognitive engagement occurs when students are asking questions, applying knowledge in new ways, and self-assessing their learning. The STEM practices reflect this kind of cognitive engagement. The exploration tasks lead students to the desired learning outcomes and tie together the other components of the W.H.E.R.E. Model. These activities should address the Essential Questions in the *why* section, which in turn lead to deeper student understanding. We focus on developing the E Guidepost, along with the R Guidepost, in Chapter 5.

Strategic Guiding Questions

- What scenario will "hook" the learner and foster engagement?
- How will the students be cognitively engaged?
- Which STEM practices will the students engage in?
- How will the explorations provide evidence of this engagement?
- How do the explorations address the Driving Question?

EXCITE: What is the hook to excite the learner?	**ENGAGE:** How will the students be cognitively engaged throughout the unit?	**EXPLORE:** What activities will help students address the Driving Question?
Create the scenario to engage the learner.	List the STEM practices that will be used as evidence.	List questions for students to investigate that will lead them to a deeper understanding of the content and skills.

FIGURE 2.6

As we explore Chapters 3, 4, and 5, each of the guideposts will be described in greater detail through the lens of three different STEM unit examples. The template is a framework and, as you will see in each of the examples, there is no one "right way" to record the information that is important for your STEM unit. Each chapter provides a slightly different approach. The completed STEM Lesson Guidepost Planning Template for each unit can be found in the appendices.

Using the Planning Template

The STEM Lesson Guidepost Planning Template is not designed as an exercise to complete from top to bottom, but rather as a place mat on which you can record and organize your thoughts during the process of brainstorming as you develop your own STEM unit. Over the next few chapters, you will see how this template is used to plan units, and then, in Chapter 7, you'll have the opportunity to try it for yourself. By breaking apart the components of the W.H.E.R.E. Model and using different grade-level and subject content examples, we can better showcase the process for completing the template. It's important to remember, however, that the template needs to be considered as a whole, rather than just focusing on one section at a time.

Think About It

Through engaging STEM units, students will learn the core concepts and big ideas using an approach in which they apply their knowledge in engaging and relevant ways. We hope that you are now beginning to see W.H.E.R.E. we are going as we begin our learning journey together.

Speed Bumps

- Why was the W.H.E.R.E. Model developed?
- What are the five essential building blocks for developing the W.H.E.R.E. Model?
- What is the key research used in the development of the STEM lesson guideposts?
- How does the analogy of the guidepost reflect the planning model of UbD?
- How should the STEM Lesson Guidepost Planning Template be completed?

The *W*—Picking the Destination

What Needs to Be Learned and Why?

Alice came to a fork in the road. "Which road do I take?" she asked. "Where do you want to go?" responded the Cheshire cat. "I don't know," Alice answered. "Then," said the cat, "it doesn't really matter."

(Carroll 1946)

As the Cheshire cat points out, you need to know where you are going before you start your journey. In the W.H.E.R.E. Model, the first guidepost focuses on the *W*, the *what* and the *why*, to help you find that destination.

What includes the "big ideas"; in other words, the key concepts, knowledge, and skills that the students will need to know and be able to do as a result of the learning. These are likely to be based on the standards you teach. The fundamental question is, what are the desired results or learning goals for the instruction? The second part of this guidepost is why: Why would the students care about learning this concept, knowledge, or skill? The why helps us think about the purpose for the learning and frame it in a way that makes it appealing and motivating to the students.

Much of this guidepost focuses on determining the Essential Questions that focus the instruction, acting as beacons to keep the learning on the right route, aimed at the ultimate destination. Then we create a single Driving Question that ties together and provides for application of the content, knowledge, and skills. Clearly articulating what needs to be learned and why it should be learned will establish the foundation for the development of the overall integrated STEM learning experience.

As a reminder, here are the strategic guiding questions for this guidepost:

- What are the desired results, including the "big ideas," content standards, knowledge, and skills (e.g., twenty-first century skills) that this lesson/unit will teach?
- Why would the students need to know and understand these concepts and skills?
- What are the Essential Questions that will lead to student understanding of the content?
- What is the Driving Question that will lead to the development of the integrated task that provides the application of the content, knowledge, and skills?

In this chapter, we'll emphasize this first guidepost as we look at creating a second-grade unit titled *Engineering a Hat*. This unit focuses on the engineering design process, but draws on other content areas as well. The students explore weather and the geographic influences on it, along with geometry, as they design a hat to wear in a particular place. In the unit the students are going on an imaginary trip and are going to need a hat. To engineer the right kind of design for their hat, they will need to learn about the geography and weather conditions where they are going. They will use geometric shapes to build their hat. A completed STEM Lesson Guidepost Planning Template for the unit is provided in Appendix C. (We'll be doing the same with other units in the chapters that follow.) As we noted in the previous chapter, the template needs to be considered holistically because the guideposts are interrelated. We focus on the W for this unit, but will address the other guideposts as well.

Beginning the Learning Journey with the What and the Why

Let's start by examining *what* we need to cover.

Deciding on the Driver

Keeping with our journey analogy, every STEM unit will have one, or at the most two, driving content standards leading the instruction. Choosing the driver for a particular unit need not be the burden of a single teacher but can be the collaborative efforts of a team of teachers or a group of district curriculum leaders. In the following example, the emphasis for the district's overall yearly goal came under the overarching umbrella of engineering design.

To choose the driver for the *Engineering a Hat* unit, we start by considering the engineering design process as described in the Next Generation Science Standards (NGSS). The Engineering, Technology, and Applications of Science (ETS) standards (Achieve Inc. 2013) gather the related engineering processes into three broad categories:

- ETS-1: Defining and Delimiting Engineering Problems
- ETS-2: Developing Possible Solutions
- ETS-3: Optimizing the Design Solutions

Together, these three categories form the primary focus of the unit's instruction. In the elementary grades, it's difficult to separate and focus on only one of the engineering/design standards. It is not that the individual learning objectives of the standards cannot be isolated, but that it is difficult to cut short the learning experience when student excitement is at its peak. What fun is it to talk about a problem or generate a solution without the opportunity to try building it? How do you foster persistence, experimentation, and self-reflection if you don't allow time for solution improvements? Thus we group the three standards together and consider all three of them to be the driver. For older students with more content knowledge, life experiences, and better research skills, the individual standards can be more readily isolated into discrete learning goals that can act as independent drivers in their own STEM units. (See Figure 3.1.)

In this particular unit, students first gather information about a situation that people want to change to define a simple problem that can be solved through the development of a new or improved object. Then they will develop a physical model to illustrate how the object helps it function as needed to solve that identified problem.

Engineering/Design Standards (NGSS)*

Standard	In Grades K–2	In Grades 3–5
ETS-1: Define	Ask questions, make observations, and gather information about a situation people want to change to *define a simple problem* that can be solved through the development of a new or improved object or tool.	*Define a simple design problem* reflecting a need or a want that includes specified criteria for success and constraints on materials, time, or cost.
ETS-2: Develop	*Develop a simple sketch, drawing, or physical model* to illustrate how the shape of an object helps it function as needed to solve a given problem.	*Generate and compare multiple possible solutions* to a problem based on how well each is likely to meet the criteria and constraints of the problem.
ETS-3: Optimize	Analyze data from tests of two objects designed to solve the same problem to *compare the strengths and weaknesses of how each performs.*	Plan and carry out fair tests in which variables are controlled and failure points are considered to *identify aspects of a model or prototype that can be improved.*

*(NGSS Lead States. 2013. Next Generation Science Standards: For States, By States. Washington, DC: The National Academies Press.)

FIGURE 3.1 Adapted from the NGSS

They will compare their solutions to others to see how it performs for the resolution of the problem.

The Passengers: Additional Content and Skills

Along with the "driver," there can also be other supporting standards as "passengers" in the STEM "car." Selecting those passengers is part of your decision making as you begin to think about your overall STEM unit. Looking over the other standards you teach can help you think of a creative way to tie the standards together that promises a unique or engaging premise for the learning.

In this case, we have already determined that the engineering design process is the main driver. Now we will examine the grade 2 content from the other disciplines such as science, math, and social studies. We notice, for instance, a science standard related to understanding weather patterns and changing weather conditions. There is a geometry standard in math that students identify geometric shapes and explore equal shares of a whole that may not be the same shape. In social studies, a portion of the student learning is focused on learning about geography, where students become familiar with the physical features of different landforms, mountains, rivers, deserts, plains, and so on. Can we bring these together to develop a relevant scenario that would lead us to the development of the Driving Question?

Setting the Purpose: The Why

Before we can develop that Driving Question, we need to consider the second part of the W Guidepost, the why. This addresses not only the issue of student motivation ("Why do I need to know this?"), but also the teacher's motivation ("Why am I teaching this?"). Focusing on Essential Questions can help identify the why for a STEM unit. In the case of *Engineering a Hat*, the Essential Questions might include:

Engineering Design

- What observations and questions need to be answered before you can solve the problem?
- How can the engineering design process be used to find a solution to the problem?
- How will drawing a picture or creating a blueprint of the design help?
- Is there more than one design that will provide for the best solution to the problem?

Science

- How can we find out about the weather where we live?
- What is the best tool for measuring the temperature?
- Why is there a difference between the nighttime and daytime temperature?

Mathematics

- How can we use shapes?
- How can shapes be divided into equal parts?
- What are the attributes of different shapes? How are they alike? How are they different?

Social Studies

- How do we describe our landscape?
- How might the land features affect the climate of an area?
- Why are there differences in temperature in an area?

Putting It Together: The Scenario and the Driving Question

Drawing on these Essential Questions, can we develop a scenario and a Driving Question that will bring together all this content? To make the learning process interesting, we need a problem that students can easily relate to before they try to engineer a solution—an engaging scenario or storyline to help them become active participants. The Driving Question captures and communicates the purpose of the learning by providing a real-world context in which students are given the opportunity to apply these key understandings.

Scenario

The scenario is not finalized until we reach the second E Guidepost, but it is important to start thinking about it now as we draft the Driving Question. In the case of *Engineering a Hat*, we drew on all the content standards and developed this scenario:

> *The students will imagine that they are going on a summer driving trip with their family from Phoenix to the Grand Canyon. While visiting the Grand Canyon, they will have a few days to walk around the rim of the canyon and perhaps even hike partway down one of the trails. They are very excited and need to gather some information for their visit. First of all, they will need to know what the weather is going to be during their stay. Then they will need to come up with the right kind of hat to protect them during the trip.*

To design their hat, students will need to incorporate some of the math standards for geometric shapes. They will need to find out about the geography and weather of the place they are going to visit.

Based on this scenario, we can revise our Essential Questions to make them more specific to our unit. This kind of revision happens often as we make our way through the template; each section informs the others, and so we are constantly refining our plans as we go.

Engineering Design

- What observations and questions need to be answered before you can solve the problem?
- How can the engineering design process be used to find a solution to the problem?
- How will drawing a picture or creating a blueprint of the design help with the construction of our hats?
- Is there more than one design that will provide for the best solution to the problem?

Science

- How can we find out about the weather at the Grand Canyon?
- What is the best tool for measuring the temperature?
- Why is there a difference between the nighttime and daytime temperature at the Grand Canyon?

Mathematics

- What shapes can be used to make our hat?
- How can shapes be divided into equal parts to make our hat look even?
- What are the attributes of the different shapes that are used in my hat?

Social Studies

- What is the landscape like at the Grand Canyon?
- How do the land features affect the climate of the Grand Canyon?
- Why are there differences in temperature between the top and bottom of the Grand Canyon?

THE DRIVING QUESTION

Now, with a scenario in mind and a good sense of our Essential Questions, we can create a Driving Question to tie the development of the integrated tasks together in such a way as to provide for the purposeful application of the content, knowledge, and skills as identified in the "What" section of the template. For the *Engineering a Hat* STEM unit, the Driving Question would read as follows:

> *How can we, as engineers, design a hat that will protect us from the hot sun and also keep us cool?*

Notice how the question articulates the work the students will do by identifying what role they will act in, what task they will do, and what problem they are trying to solve. It is short, direct, and specific, yet presents an open-ended challenge that produces a variety of results.

Engineering a Hat: Template

In Figure 3.2, you will see the completed "What/Why" section of the STEM Lesson Guidepost Planning Template for the *Engineering a Hat* unit.

WHAT are the desired results, including big ideas, content standards, knowledge, and skills?

List the content standards and what the students will know and be able to do.

Students will design a hat that can be worn in the hot, sunny conditions but also have a way to keep them cool. They will compare similar designs, describe the shapes used, and discuss the advantages/disadvantages of the design for the type of weather conditions it will be used for.

MAIN STANDARD (Driver)

Engineering Design

ETS-1: Defining and delimiting engineering problems
ETS-2: Developing possible solutions
ETS-3: Optimizing the design solutions

SECONDARY STANDARDS (Passengers)

Science: ESS2.D: K–2 Weather and Climate

Weather is the combination of sunlight, wind, snow or rain, and temperature in a particular region at a particular time. People measure these conditions to describe and record weather and to notice patterns over time.

Social Studies Standard: Strand 4: Geography
PO3: Discuss physical features (e.g., mountains, rivers, deserts) in the world.

(With a connection to science: measure and record weather conditions, identify clouds and analyze their relationship to temperature and weather patterns.)

Mathematics
Geometry, reason with shapes and their attributes

2.G.A.3: Partition circles and rectangles into two, three, or four equal shares using the words halves, thirds, half of, third of, and so on and describe the whole as two halves, three thirds, four fourths. Recognize that equal shares of identical wholes need not have the same shape.

WHY would the students care about this knowledge and these skills?

Craft the Driving Question that will lead to the development of the integrated tasks that provide for the application of the content, knowledge, and skills.

List the Essential Questions that can be answered as a result of the learning.

DRIVING QUESTION
How can we, as engineers, design a hat that will protect us from the hot sun and also keep us cool?

ESSENTIAL QUESTIONS

Engineering Design

- What observations and questions need to be answered before you can solve the problem?
- How can the engineering design process be used to find a solution to the problem?
- How will drawing a picture or creating a blueprint of the design help with the construction of the hat?
- Is there more than one design that will provide for the best solution to the problem?

(continued)

FIGURE 3.2 Summary of Guidepost

Science
- How can we find out about the weather at the Grand Canyon?
- What is the best tool for measuring the temperature?
- Why is there a difference between the nighttime and daytime temperature at the Grand Canyon?

Mathematics
- What shapes can be used to make our hat?
- How can shapes be divided into equal parts to make our hat look even?
- What are the attributes of the different shapes that are used in my hat?

Social Studies
- What is the landscape like at the Grand Canyon?
- How do the land features affect the climate of the Grand Canyon?
- Why are there differences in temperature between the top and bottom of the Grand Canyon?

FIGURE 3.2 *(continued)*

Engaging the Students Using the "Needs to Know"

As we wrap up our discussion around the W Guidepost, our attention turns to how to present the scenario and the Driving Question to the students. We've found it effective to begin by brainstorming the "needs to know" questions they have, information they think they need, and anything else they think they need to know to answer the Driving Question. This brainstorming session helps them to understand the type of design they need to consider, what kinds of research they will need to do, and what other parameters might be needed to help keep them on track to successfully complete the task. It can also reveal misunderstandings of the task early. It will encourage further engagement and buy-in because the students will have had a chance to clarify what they are expected to do. It's important to list the needs to know questions so students can see them and discuss which are pertinent to helping them find solutions to the Driving Question (not all of them will be). The discussion helps students understand the process of identifying what information is important or useful. It puts the students in the position of making decisions on how to approach the problem (though you may need to ask some leading questions to guide their thinking).

Some suggested prompts to begin the discussion of the needs to know questions are:

> To engineer your hat to fit the location of our hike, what research do you think we will need to do?
>
> Where is the Grand Canyon? What does it look like around the rim? Are there trails we can follow?
>
> What do you think the weather will be like in the Grand Canyon during our visit?

How can we find out what the temperature will be during our visits?

What do you think the daytime temperature will be? The nighttime temperature?

What do you think will be a good design for the hat?

How can we make sure the hat will cover both sides of our face?

Could we engineer our hats to both protect us from the sun and also keep us cool?

And it's important to help students begin to think of limits to their solution ideas. These last questions can help set up the constraints that the students will need to consider and might include:

How much time do we have to work on our hats?

Can we work with a partner?

What materials will we be using?

Do we need to draw our design blueprint first before we begin to construct our hat?

Summarizing the Other Guideposts—H.E.R.E.

The H Guidepost provides an opportunity to begin to lay out the route that you will take to reach the destination. What are the major highlights that would be important as the student's progress toward the goal? In the *Engineering a Hat* unit, the sequence of student experiences moves from introducing the scenario, Driving Question, and engineering design process to eliciting the needs to know questions, doing the research, creating their blueprint, constructing their hats, and finally students presenting their solutions. Each of these in turn could provide points to assess student learning, which leads to the next guidepost—evaluation.

The E Guidepost provides an opportunity to consider what success will look like. How will students demonstrate their new understandings? The summative section gives a checklist of important criteria for success. Does their design solve the stated problem? Does it include equal parts of different shapes? How did they refine their design? Can they describe the land features and weather conditions that make their design effective? The major milestones in the H Guidepost can help identify interim assessments, monitoring student progress along the way. For example, can they identify the major landforms and features of the Grand Canyon? Can they describe the weather conditions at the Grand Canyon for the time of year they chose? Are they able to partition shapes appropriately? All of these ensure students are attaining the individual goals for all the content identified in the W Guidepost. Considerations should also be given to what prerequisite skills are necessary to initiate the student explorations. Do they know their shapes? Can they partition them into equal parts? Knowledge of certain vocabulary can facilitate learning, so making sure they all have a common language to share their ideas effectively is part of that preassessment development.

The R Guidepost provides an opportunity to make sure the tasks are sufficiently complex and intriguing to make the students active participants and not just doers of busywork. In the *Engineering a Hat* unit, having students choose another destination for their summer trip would result in a greater variety of hat designs and let them compare designs from different places. Having students make choices moves the learning to a more rigorous experience. To improve relevance, the students can learn about hats from different places in the world, or perhaps select their favorite hat and tell why. Perhaps having the students choose a different article of clothing to design would make it more personal to them. Inviting a clothing designer/engineer in to talk about how they design an article of clothing can connect students' classroom experience to the real world.

A Note About Technology

You may ask, "Where is the *T* in this STEM unit?" Remember that technology is not just a device or a plug-in. Although students may have used plug-in technology to do their research, the broader definition of technology is more apparent in the work that they did. Keep in mind that technology includes all the ways in which humans have changed their world to meet and achieve their needs and goals. The students used different tools, such as scissors, tape, rulers, and measuring tape, to construct their hats.

The final guidepost, E Guidepost, provides an opportunity to add detail to the initial scenario, developing it further so that it can capture the overall intent of the transdisciplinary task and relate back to the Driving Question. Also, as the students become involved in the work, the questions become, "How will they be cognitively challenged to remain mentally engaged in the task?" and, relatedly, "How can students engage in the STEM practices?" In constructing their hats, the students will focus on developing and using models, while their explanations of their hat designs allow them to construct explanations and design solutions and engage in argument from evidence as they defend their design solutions during the peer review process. Their choice of materials, tools, and approach emphasizes "use appropriate tools strategically."

Finally, the "Explore" section gives a place to summarize the details of the activities outlined in the H Guidepost. What are the materials, strategies, and organization that will be necessary to successfully implement this learning journey? What are the rules of the classroom for working together? Here, some materials might include: straws, paper plates, yarn, tape, measuring tapes, scissors, cotton balls, different pieces of fabric, string, ribbons, extra scraps of paper. You might set norms for working together such as, "If called upon, any member of the team must be able to fully explain the design and practicality of the group's hat."

A full STEM Lesson Guidepost Planning Template for this unit is in Appendix C.

Think About It

As the Cheshire Cat said to Alice, "If you don't know where you are going, then any road will take you there." Keep in mind as you begin to develop your own STEM units: always start with "*What* do you want the students to know and understand?" And then ask yourself, "*Why* are they learning this material and why would they care about this learning journey?" Look at what you are already teaching and think about how you might be able to craft a creative scenario and a Driving Question that connect a series of integrated activities together into a richer learning experience for the students. Now, let's continue on this learning journey together as we examine the "H" and "E" sections of the template.

Speed Bumps

- Where can you address needed skill instruction in the template?
- How might the specific content that you need to cover in the curriculum be applied to the template?
- How might you modify or change the Essential Questions to better support the content?
- How does altering the current Essential Questions impact the focus of the Driving Question?

The *H* and *E*—Planning and Evaluating the Route

If you don't know where you are going, you might wind up someplace else.

—Yogi Berra, 35 of Yogi Berra's Most Memorable Quotes (*New York Post* 2015)

In this chapter, we will explore the development of the *H* (how) and the first *E* (evidence and evaluate) guideposts in the W.H.E.R.E. Model. To plan any learning experience, it is first important to determine what you want students to learn (which we discussed in the previous chapter), then determine how you will assess that learning, and finally, plan the instruction and learning experiences that will get students there. What the students will do, how they will do it, and in what order they will do it in are essential to their achieving the outcomes you expect. Thinking about the path helps you avoid a destination that is unattainable, or in Yogi Berra's words, "ending up someplace else." The *H* and the first *E* are therefore connected because the destination (evaluation) must reflect the progression of experiences the student will engage in throughout the unit.

The example presented in this chapter is built around an energy unit based on middle school science standards. Students will build a wind energy turbine to investigate energy transformations. Before we look at these new guideposts, let's take a brief look at the *W*s for this unit: the *what* and *why*.

Summarizing the *W*: What and Why

As described by the National Research Council in *Developing Assessments for the Next Generation Science Standards*, "The kind of instruction that will be effective in teaching science in the way the framework and the NGSS envision will require students to engage in scientific and engineering practices in the context of the disciplinary core ideas—and to make connections across topics through the cross-cutting ideas" (2014, 4). The *Wind Turbine* unit embodies the kind of instruction envisioned in the new science standards.

This STEM unit is an engaging investigation because it allows students the opportunity to explore ideas and to be actively involved in the design, building, and manipulation of various factors that impact the energy output of a wind turbine system. Groups of two or three students work together. The small-group structure results in more designs to compare and greater amounts of data to evaluate at the conclusion of the unit.

What

Although this unit could easily be considered an engineering/design unit, we chose science as the driver for this example to show how the selection of the primary focus influences the direction of the investigations within the unit. With science, the emphasis is on developing an understanding of the relationship between changes in kinetic energy and the energy output of a system. Therefore, what will be assessed (*E*) and the progressions of learning (*H*) to reach that destination are all focused on that core understanding.

The driver of this unit is a science standard (Next Generation Science Standard [NGSS]: MS-PS3-5), which states, "Construct, use and present arguments to support the claim that when kinetic energy of an object changes, energy is transferred to or from that object." The main focus of the student work in this unit is to develop and solidify students' understanding of two core ideas: (1) when there is a change in motion of an object there is a change in energy, and (2) when two objects interact, each one exerts a force on the other such that energy can be transferred to and from the objects. Using the kinetic energy of the wind (produced by a window fan), students will observe that the force of the wind moves the wind turbine blades and that the energy of the wind is transferred to the blades, causing them to turn. This motion is converted into electrical energy in the motor that can then be measured as electrical output. The students will investigate different ways to increase the motion of the blades from a given force and use those changes in electrical output as evidence of the change in the kinetic energy of the blades. They will then make and defend a claim about which blade conditions transfer the greatest amount of kinetic energy.

To achieve this understanding, students will also be involved in the engineering design process, developing models for iterative testing to propose an optimal design solution. Students will also apply several key mathematical ideas as they engineer their designs, that is, they will collect and organize data as their evidence. They will need to analyze the quantitative relationships between dependent and independent variables as they manipulate their assigned design parameter for the turbine blades. Although both the engineering process and the mathematical analysis work are important, they are not the primary focus of the unit. They are two of the passengers on this STEM journey.

English language arts (ELA) content also plays a role in this unit. In the development of the scenario, the final product includes both a group visual presentation of the evidence to support the students' claims and an individual written report describing their reasoning in defense of that evidence. Adding two writing standards was a natural extension of the unit's final product and thereby ELA becomes another passenger.

Why

The core understanding of the unit investigation can be summed up in the following statement: "Provide evidence as to which blade design in the model wind turbine transfers the greatest amount of kinetic energy." This captures the work that the students must do and sets up the need to state a claim about their chosen blade design and gather evidence to support that claim. As we started creating this unit, the Driving Question (focused on the core science understandings) originally was "Which blade design in the model wind turbine transfers the greatest amount of kinetic energy?" But as we began thinking about a real-world context to use as a scenario, we saw the need to revise the Driving Question. As you read the scenario, think of how the Driving Question could be revised to incorporate the real world setting.

Scenario

The town is soliciting bids for an efficient wind energy generator built on a research-based design. The generator will be placed on the hill in the center of town, which gets a constant flow of westerly wind. Present a proposal for an efficient wind turbine design based on research. Contract award will be based on design research and committee presentation.

Here is our revised Driving Question, linking the content with this context.

REVISED DRIVING QUESTION

As the Wind Turbine Power Company representative, present evidence that addresses the question, "Which blade design transfers the greatest amount of kinetic energy?"

The Essential Questions are geared toward supporting this Driving Question and reflect the development of student understandings. They provide key landmarks that guide the progress of the unit. As students formulate their answer to the Driving Question, they will be reminded to reflect on the Essential Questions that support it. In the initial outline, there were only three EQs, which reflected the science content understandings: "How can kinetic energy in a system be measured?"; "How can the kinetic energy in a system be improved?"; and "How is kinetic energy transferred between objects or systems?" We added five additional EQs to address the passengers, as shown in Figure 4.1, which summarizes the W Guidepost for this unit.

Notice that the Essential Questions are organized in such a way as to step the students through the learning experience. This order comes as a result of the H Guidepost, to which we now turn.

H: How Do I Plan to Meet This Goal?

Having identified the standards, how do you begin to craft the learning progression for the unit? There are generally some key sequences in the learning for the unit investigation to progress in a meaningful and systematic way. The H Guidepost provides assistance in unraveling those key learning progressions, which in turn support the development of the summative assessment tool in the E Guidepost. As you begin, think about the following strategic questions:

- What learning experiences will enable the students to understand the concepts and skills in meaningful ways?

WHAT are the desired results, including big ideas, content standards, knowledge, and skills?

List the content standards and what the students will know and be able to do.

The students develop an understanding of the relationship between changes in kinetic energy and the energy output of a system. They discover that the electrical output can be manipulated by design changes in the system.

By developing a model wind turbine to transfer electrical output data, students produce evidence that changes in kinetic energy are transferred from one system to another.

MAIN STANDARD (Driver)

MS-PS3-5: Construct, use, and present arguments to support the claim that when kinetic energy of an object changes, energy is transferred to or from that object.
DCI #1: When the motion of an object changes, there is inevitably some other change in energy at the same time.
DCI #2: When two objects interact, each one exerts a force on the other that can cause energy to be transferred to or from the object.

SECONDARY STANDARDS (Passengers)

Engineering/Design

ETS1-4: Develop a model to generate data for iterative testing and modification of a proposed object, tool, or process such that an optimal design can be achieved.

ELA

W.6.1.A: Introduce claim(s) and organize the reasons and evidence clearly.
W.6.4: Produce clear and coherent writing in which the development, organization, and style are appropriate to task, purpose, and audience.

Mathematics

6.EE: Represent and analyze quantitative relationships between dependent and independent variables.
6.G: Solve real-world problems involving area, surface area, and volume.

WHY would the students care about this knowledge and these skills?

Craft the Driving Question that will lead to the development of the integrated tasks that provide for the application of the content, knowledge, and skills.

List the Essential Questions that can be answered as a result of the learning.

DRIVING QUESTION

As the Wind Turbine Power Company representative, present evidence that addresses the question, "Which blade design transfers the greatest amount of kinetic energy?"

ESSENTIAL QUESTIONS

Engineering Design

- How is kinetic energy transferred between objects or systems?
- How can kinetic energy in a system be measured?
- How can the kinetic energy in a system be improved?
- How does the blade configuration, length, material, or angle impact the amount of kinetic energy transferred?
- How can the surface area of the blades be calculated?
- How do changes in the model wind turbine design affect its electrical output?
- How can I use the data I generate to support my claim?

FIGURE 4.1 Summary of the W Guidepost

- How will the sequence of these learning experiences help students to construct their understanding and apply the skills?
- What are the major tasks or milestones that will lead toward answering the Driving Question?

Note that this section is not about specifying all the details of each experience; we will address those in the next chapter in the second E Guidepost. The focus here is on organizing the experiences in a logical way to scaffold student understanding. If we think of the ultimate understanding as a destination, we need to think through what core activities will lead the students to that destination.

In this unit, the intended outcome of the science standard is to have students "construct, use, and present arguments to support the claim that when the kinetic energy of an object changes, energy is transferred to or from that object." The construction of the wind turbine provides opportunities for investigating the various parameters of blade design that affect changes in kinetic energy. The introductory activities in the unit should initially focus on confirming students' basic content understanding and core knowledge. These activities should include building common experiences so all students have an equal opportunity for success, including developing a uniform vocabulary that ensures consistency as ideas are shared among the students. Because the major focus of this unit is on the science standards, not the engineering or construction process, and because the wind turbine kits were new to the students, we decided to introduce an initial turbine construction activity as an exploratory endeavor. Students would work in small teams and construct similar wind turbine models so that all could develop familiarity with the turbine structure, understand its operation, and explore the features that will become the variables in the future investigations. They will construct the wind turbine and understand how it generates the electrical output from the kinetic energy of the wind. The intent here is to get the students familiar with the wind turbine system. The development of the first two Essential Questions—"How is kinetic energy transferred between objects or systems?" and "How can kinetic energy in a system be measured?"—correspond to this sequence of initial activities in the *Wind Turbine* unit.

Once students have this basic familiarity, then the student teams can work on more open-ended tasks such as investigating which blade variables produce the greatest change, gathering specific energy output data for each of their trials. This is core to the science learning objectives where students work on manipulating the different design variables.

Each team is assigned a specific feature to investigate (the number of blades, the length of the blades, the material of the fin on the blade, or the angle of the fin on the blade in relation to the hub). This decision was based primarily on class time constraints, minimizing the number of variables and design trials the students would have to deal with, and allowing them to focus their attention on the limits of a given parameter. Having several teams working on each feature provides opportunities for comparison of results and more in-depth discussion afterward on procedures and processes. For efficiency, the student teams will have limited time to explore just one of the blade variables.

As students work, you'll have an opportunity to observe and give students extra support as needed. Students will measure the changes in kinetic energy as the electrical output of the wind turbine system as they manipulate these design features. The electrical output is the result of the motion of the turbine blades turning the motor. As the motion of the blades change, the kinetic energy changes, and so does the electrical output. Students begin to connect an increase or decrease in the speed of the blades because of changes to the blade design with the corresponding changes to electrical output. This ties to the next two EQs: "How can the kinetic energy in a system be improved?" and "How does the blade configuration, length, material, or angle impact the amount of kinetic energy transferred?" Together, they focus on the engineering design process.

The overall goal is to generate data about the relationship between changes made to the blades and the energy output of the system. The questions "How do changes in the model wind turbine design affect its electrical output?" and "How can the surface area of the blades be calculated?" will require students to know how to measure surface area of the blades as a technical aspect of their designs.

Both of these questions also give students direction for undertaking further exploratory testing, determining what other variables they might want to investigate as their natural inquisitiveness moves the investigation forward. They might want to try combining different design parameters together, like shorter blades with a given fin angle of a certain material. Or they may decide to alternate or combine blade materials to see if a variety of materials work better than just one type.

The students will then consider the following question: "How can I use the data I generate to support my claim?" This connects to the final student product where they will report their data as evidence to support the claim that their wind turbine design will generate the most electrical energy for the town.

Our proposed learning sequence is detailed in the "H" section in Figure 4.2. The sequence assumes five to seven forty-five-minute class periods. This was based on our experience with the turbine construction and having the blade materials prepared ahead of time. To facilitate the management of the unit, it is important to make sure that students have a clear understanding of the purpose, the time frame, and the expectations. We propose that you review the initial guidelines for the turbine challenge, discuss the investigation constraints, and review the criteria on the evaluation rubric as part of the planning process. This saves a lot of repeating directions to each individual team as the investigation unfolds. Consider your own schedule, your student needs and experience, and your comfort level with the activities as you plan the sequence of activities.

The H Guidepost guides the initial development of the sequence of student learning experiences. The details of that sequence evolve as more planning details are added later in the template. In particular, as we begin to define our assessments, we can see how they can impact our proposed pathway, making further modifications necessary.

HOW do I plan to meet this goal?

Identify the pathway, including major tasks and milestones that result in answering the Driving Question.

1. Small groups of students will work on constructing the turbine first, before more thorough investigation on given parameters. Will assess student proficiency after first class, and regroup student teams as necessary.
2. Review the initial guidelines for the turbine challenge, discuss the investigation constraints, and review the criteria on the evaluation rubric.
3. Three milestones: initial construction success; generation of the team's "variable" data; final presentation.
4. Proposed five class periods to accomplish:
   Class 1: Build turbine and make modifications.
   Classes 2–4: Small-group investigations of given parameters; finalize report.
   Class 5–7: Present oral and written reports.

FIGURE 4.2 Summary of the H Guidepost

E (Evaluate): What Evidence of Learning Will Be Used and How Will I Evaluate the Final Product?

Although assessment may seem like a culminating activity, it should be viewed as the primary target by which the students demonstrate their understanding of what they were expected to learn. Planning assessment requires reflecting on what you want the students to know (the *W*), and deciding what evidence you want them to produce to demonstrate that acquisition of new knowledge. The following strategic questions reflect not only summative assessment (the final product), but also the importance of assessing student understanding all along the journey:

- What prior knowledge and skills will the students need to demonstrate?
- What formative assessments will be used to measure student progress toward understanding and to inform instruction?
- What summative assessment, culminating product or task(s), will demonstrate student understanding of the standards, content, and skills?
- What criteria and tool(s) will be used for assessing student success?

The Final Product: Summative Assessment

For the *Wind Turbine* unit, the culminating product must be crafted in a way so that the students can engage in argument over the claim they make, using evidence they produce to support that claim. The claim must be about the change in the transference of energy in a system when the kinetic energy changes. The development of argument from evidence cuts across many of the content disciplines, and is an important life skill as well. It particularly reflects ELA standards as well as the NGSS. We chose a presentation and report based on their investigation data to reflect real-world uses of these skills and provide an authentic purpose for students' writing. The purpose is set as the presentation to the town council describing their turbine design and its supporting research data.

The resulting summative product has two parts, one representing the group's collective experience and one focused on individual accountability for the student's own understanding. Including a group presentation holds all group members responsible for having reliable and accurate data and a sound argument. The group works together to produce an oral and visual presentation sharing the details of what they have learned: their procedure, their results, and their conclusion. In the second, individual part of the unit assessment, each member of the group must write their own one-page summary addressing the original claim, "In the wind turbine model, which blade design transfers the greatest amount of kinetic energy?" and citing the evidence gathered from her investigation and detailing her rationale for selecting that evidence.

The two products are complementary. The group assessment task focuses on the organization of the evidence and ties directly to the science standard (our driver for this unit). It also addresses the twenty-first-century skills of collaboration, communication, critical thinking, and creativity. The individual report further assesses the content understanding that is central to the science standard and also affords an opportunity to add the passengers, particularly the ELA writing standards.

Developing Scoring Criteria

Scoring rubrics are a useful tool, not only for assessing the student product or performance, but also to help students better understand what the parameters for success look like. The rubric defines the criteria on which the product or performance will be evaluated and should contain descriptive guidelines for the completeness of the work. Generally a rubric of three to six criteria works best so as to have enough criteria to consider the multiple facets of the student work without overwhelming or confusing students.

In the *Wind Turbine* unit, the student work should exhibit strong reasoning about the relationship between changes in kinetic energy and the energy transfer in the system. The claim students make regarding this relationship should be clearly stated. As the students change the conditions of the turbine system, the subsequent change in electrical energy output represents the data to be used as their evidence, either supporting or refuting their claim, and this should also be reflected in the rubric. The criteria should capture the desired completeness of the work so students can use it as a tool to self-monitor their own submission. The addition of the report quality criteria is designed to shift the scoring emphasis away from technical errors or omissions that result from poor writing habits and allow the teacher to focus on science conceptual understanding. See Figure 4.3 for a rubric for this unit.

You might also want a separate rubric to assess skills such as brainstorming, teamwork, and collaboration, separating the goals of fostering positive social interactions from the task of developing conceptual understanding. The sample rubric in Figure 4.4 gives space for students to self-assess and for the teacher to provide comments as well. A follow-up conference allows you to discuss and reinforce positive social behaviors.

Note that the two rubrics in Figures 4.3 and 4.4 have different scoring levels. We did this to highlight that scoring is always somewhat subjective, and it is important for everyone involved to agree. The choice of three or four levels is somewhat arbitrary and may depend on your school or district policies. Whatever the number of levels, the descriptions should be based on a continuum, clearly distinguishing one level of success from the next.

Student Rubric—*Wind Turbine* Unit

Criteria	Below Proficient	Proficient	Above Proficient
Claim	Claim is missing or poorly stated. *And* Claim lacks scientific basis that reflects the relationship between changes in kinetic energy and the energy transferred in the system. *And* Claim does not reference energy output data.	Claim is clearly stated and mostly scientific. *And* Claim makes some references to the relationship between the changes in kinetic energy and the energy transferred in the system. *And* Claim makes some reference to the energy output data as supporting evidence.	Claim is complete, clearly stated, and scientifically accurate. *And* Claim has specific references to the relationship between the changes in kinetic energy and the energy transferred in the system. *And* Claim makes explicit references to energy output data as supporting evidence.
Evidence	The data are not provided or are incomplete and do not support the claim. *And* The procedure description is minimal. *And* The procedure description lacks details about the purpose for the collection of the energy output data.	The data are complete, are mostly accurate, and mostly support the claim. *And* The procedure description provides many details of the student work. *And* The procedure description provides some details about the purpose for the collection of the energy output data.	The data are complete, are accurate, and fully support the claim. *And* The procedure description is complete and fully details the student work. *And* The procedure description provides the specific details about the purpose for the collection of the energy output data.
Reasoning	The reasoning is missing or inaccurately explains how the energy output data are used as evidence to support the claim. *And* The evidence provided shows little understanding of the relationship between the change in kinetic energy and the transfer of energy.	The reasoning is missing key details in explaining how the energy output data are used as evidence to support the claim *And* The evidence provided shows understanding of the relationship between the change in kinetic energy and the transfer of energy.	The reasoning fully explains how the energy output data are used as evidence to support the claim *And* The evidence provided shows a thorough understanding of the relationship between the change in kinetic energy and the transfer of energy.
Report quality	Report is not well organized and lacks important details that communicate the reasoning. *And* It has many errors in grammar, spelling, or data.	The report is organized and has a cohesive structure that helps to communicate the reasoning. *And* It may contain some errors in grammar, spelling, or data.	The report is well organized and coherent and has a cohesive structure that clearly and fully communicates the reasoning. *And* It has no errors in grammar, spelling, or data.

FIGURE 4.3 *Wind Turbine* Rubric

Team Assessment

Criteria	Performance Levels				Student	Teacher
	4	3	2	1		
Cooperation (Cite an example)	I work well with all group members and I share the work-load equally.	I work well with most of the group but at times I do not share the workload.	I work well with the group some of the time but the other group members do most of the work.	I do not work well with the group and I do not participate in sharing the workload.		
Participation (Cite an example)	I participate fully and am always on task in the group and class.	I participate most of the time and am often on task.	I do participate but also find that I waste time a lot and have a hard time staying on task.	I do not participate with my group and most of the time am not on task.		
Listening (Cite an example)	I am attentive and listen to what my team-mates have to say before I speak or ask questions.	I listen most of the time and I try to pay atten-tion to what my teammates are saying.	I listen some of the time but I'm anxious to share with my group what I know, therefore I tend to interrupt.	I don't pay much attention to what my teammates are saying as I have my own ideas that I want to get heard.		
Feedback (Cite an example)	I give constructive feedback most of the time.	I give constructive feedback often.	I only give feed-back when I am asked directly.	I never give feedback to my teammates.		
Leadership (Cite an example)	I welcome the opportunity to take a leader-ship role and help others in my group participate.	Most of the time, I'm open to taking on a leadership role.	I can take a leadership role but would rather do it myself.	Most of the time, I prefer to be just a group member and not be in a leadership role.		
Work habits (Cite an example)	I am always on task and never need remind-ers to do the work. I encour-age my group members to do the same.	Most of the time, I am on task and seldom need to be reminded to do the work and participate.	There have been several times when I have had to be reminded by the team to stay on task.	I try but need to be reminded to stay on task and participate with my team.		

FIGURE 4.4 Team Assessment Rubric (Vasquez, Sneider, and Comer 2013)

Using a rubric makes learning outcomes clearer and provides a target for the student to aim toward, makes the assessment process a fairer and more accurate assessment of a student's ability, and focuses on the quality of the product. It is especially useful if you have multiple reviewers scoring student work because it helps reduce individual reviewer bias.

Formative Assessment

According to *How People Learn: Brain, Mind, Experience, and School*, "The roles for assessment must be expanded beyond the traditional concept of testing. The use of frequent formative assessment helps make students' thinking visible to themselves, their peers, and their teacher. This provides feedback that can guide modification and refinement in thinking" (National Research Council 2000). Although students may see their ultimate goal as creating the final product, there are other learning milestones that can and should be assessed. To determine the types of formative assessments needed, consider questions such as how you will assess understanding of both content and skills; whether to use a single measure or a series of shorter, smaller tasks for a particular topic; and how frequently you want to check students' understanding. You will also want to consider whether a given standard or topic relates to a fundamental understanding or a more minor supporting idea. Is it necessary to have evidence of student understanding for each discipline-specific concept independently, or is the successful application of the idea in the final product sufficient? Can one assessment address multiple areas? (For instance, in the *Wind Turbine* unit, the mathematics standards focus on data, and the science standard focuses on building an argument based on data; can these two be assessed together?)

In this particular unit, in addition to the informal observations being made by the teacher, the formative assessment was a series of five open-ended response questions designed to elicit details of the students' understanding of the experimental design and their ability to identify how the changes made to the blade affected the electrical output. Students were asked to review a hypothetical turbine experiment and determine if valid engineering/design processes were applied. A data set was provided and students were asked to respond to questions dealing with the interpretation of that data. The open-ended nature of these questions allows for some direct teacher questioning as part of the evaluation process. This can help the teacher to perhaps clarify student thinking and address any concerns that may have arisen during their informal observations. Intervention strategies can be added to this section of the template as well. In this unit, the formative assessments evaluate the conceptual understanding of energy transfer and reinforce understanding of the design process. They were administered after the initial builds but prior to starting the next phase of experimentation.

Preassessment

Defining the preassessment is an important aspect of the planning routine. Preassessments aid in identifying misconceptions, assist you in understanding what your students know and do not know, and provide early insight into which students might have difficulty with the concept.

> *Since learning is influenced by what learners already know and think they know, and by their view of themselves as learners, it is essential that learning experiences be designed to elicit and connect with or challenge prior knowledge and provide opportunity for interaction with people and ideas.* (Loucks-Horsley et al. 2010, 55)

In formulating the preassessments, consider what misconceptions students may already bring to a unit or investigation. It is important to think about what gaps in student knowledge or understanding could act as obstacles and impede student success. The main questions to consider as you begin planning are, "What will the students need to know and comprehend to achieve the intent of the objectives?"; "Is the content new?"; "Is it familiar but presented in a new or different context such that the students may need to be reminded of what they already know?"; "What key vocabulary or practices are necessary for success?"; and "What challenges will be presented in the unit that they have to navigate?" For help crafting these preassessments, a series of books on formative assessment probes by Page Keeley and colleagues (2005) are available from the NSTA Press. This collection of books offers guidance and support for creating assessments that can help identify commonly held misconceptions, naive ideas, or incomplete science beliefs that students hold and that can interfere with their assumption of new information.

In the *Wind Turbine* unit, the science standard, acting as the driver, dictated what knowledge students needed to be familiar with prior to beginning the unit. It is important to check student understanding of kinetic energy and how energy is transferred because it is an essential component of the standard. Because understanding evidence-based argument and the engineering design process are also essential to the student work in the unit, these understandings should also be checked. Knowing what key terms mean, such as ***kinetic energy***, ***force***, ***motion***, and ***variables***, is necessary so that common language is readily shared among the different teams.

Figure 4.5 summarizes the E Guidepost for this unit.

EVIDENCE and EVALUATE: What evidence of learning will be used and how will I evaluate the final product?

PRE-ASSESSMENT: What prior knowledge is needed for this task?	**FORMATIVE:** How will I measure student progress toward understanding?	**SUMMATIVE:** What criteria are needed for students to demonstrate understanding of the standards, content, and skills?
Identify the prerequisite skills and understandings. Probe for student understanding on the following: • energy transfer • engineering/design process. Need to know: • kinetic energy • energy transfer • motion • force • systems and subsystems • evidence • variables.	Establish the assessment tools you will use to monitor progress and inform instruction. Individual: Open response questions (five) about the engineering/design process and the evidence produced from the initial wind turbine building activity. Intervention: Monitor student success after initial build; reassign groups if necessary to accommodate for special needs (motor skills, social skills).	Create a checklist of criteria for use in a rubric. Group: Will create an eight- to twelve-slide PowerPoint detailing procedure; results; conclusion. Individual: Will write a one-page paper responding to the claim following the claim-evidence-reason format. Rubric: Two rubrics needed (refer to ELA standards for presentation/writing skills).

FIGURE 4.5 The E Guidepost

Looking Ahead—The *R* and the *E*

In the next chapter, you will see how the R and second E Guideposts provide a framework for improving student reasoning and engagement. Here we summarize those guideposts for this unit. As you read the R Guidepost notes in the completed STEM Lesson Guidepost Planning Template (Appendix D), think about ways in which the work in the *Wind Turbine* unit moves from being teacher-centric—that is, the investigation approach is directed by the teacher—to one where the students become more responsible for generating the "what next" or "what if" as the unit progresses. This is the intent of the "Rigor" section of the template. In this unit, after the initial build, the students are asked to generate a list of blade parameters to investigate instead of being given a list of changes to make in their design. The ultimate goal is for them to recognize that a possible combination of different design parameters can be even more effective in generating the electrical output from the system. To increase the relevance of the task, and to make the scenario more personal, have students learn about the career potentials in renewal energy sources. Having a wind turbine installer come and describe the work installers do and the process used for installing wind turbines can open up ideas for future career opportunities that students may not have thought about or considered. Having students research the actual bid process used by their town could tie into other social studies standards such as civics and local government, which can become additional passengers in the STEM car.

As you look at the final E Guidepost for this unit, think about the ways in which the scenario was crafted to grab the student's attention and "hook" them into the experience. As we have already described, the scenario is used to craft the Driving Question and provides a vehicle for adding other passengers to the STEM "car." This section provides the opportunity to reflect on and refine that real-world context and provides a nexus for ensuring a cohesive and coherent connection among the multiple disciplines.

Scenario

> *The town is soliciting bids for an efficient wind energy generator built on a research-based design. The generator will be placed on the hill in the center of town, which gets a constant flow of westerly wind. Present a proposal for an efficient wind turbine design based on research. Contract award will be based on design research and committee presentation.*

A key part of the second E Guidepost is considering the STEM practices (see Chapter 1) that will engage students cognitively as they work on the investigations. (We will discuss this more in the next chapter.) In the *Wind Turbine* unit, the following STEM practices are the primary focus:

- engaging in argument from evidence
- analyzing and interpreting data
- planning and carrying out investigations.

Finally, the "Explore" section of the template offers a space for further defining the activities that form the unit. This is where you can record details for the activities laid out in the "H" section, additional questions for students, materials or readings you need, and so on. In the *Wind Turbine* unit, this section includes a little of everything. In thinking about the student work, what other kinds of reflective thought questions might you need to ask students as they do their work? Included here are some starter questions/suggestions that might be useful as prompts for some teams who seem lost or confused. These "back pocket" questions can help facilitate classroom management when the investigation is in full flight.

- Do more blades make for more electrical output?
- Are longer blades more effective for transferring the kinetic energy of the wind?
- Is there a difference in energy output depending on the way the blade is angled on the spoke?
- Gather materials: oak tag, plastic bottles, window fan.
- Suggest content reading: chapter text on energy transfer.

The finished template (found in Appendix D) makes it look like a relatively simple task: moving from one section to the next, completing each part as you go, building a coherent and structured learning experience, in an organized and sequential manner. In truth, the template is anything but simple. This unit template underwent many revisions and updates. As each section of the template was added, other sections were reworked, revised, and revisited multiple times before arriving at its current state. The template was further refined after it was initially introduced with students.

> **Note:**
>
> The Wind Turbine materials used for this activity were purchased online from Teachergeek.com (888-433-5345). TeacherGeek.com has a series of inexpensive STEM kits and Maker Space modules with reusable materials that support ongoing, open explorations in many areas of the curriculum.

Think About It

Remember, the template is like a painting canvas, where, with the addition of each new color, a clearer, sharper image begins to emerge. Thinking about the destination guides your selection of colors from the palette, and the work can only be considered complete when there is sufficient detail to render the full scope of the painting. As with any work of art, it should invite interpretation and inspection. The completion of the "H" and "E" sections go hand in hand as you formulate your assessments and plan the instructional sequence.

Speed Bumps

- Can you complete the template in different sequences, especially if you adhere to the backward design planning model?
- How will you determine what the final product of the unit will be?
- How might you use the Essential Questions to help craft your scoring rubric?
- What strategies can be used to develop effective preassessment tools?

The *R* and *E*—Keeping the Trip Interesting

Integrated approaches to teaching and learning are not new. A century ago, John Dewey and his contemporaries suggested that by applying ideas from one discipline to another, students would come to appreciate the interconnection of ideas and the relevance of their schooling.

—John Wallace, et al., *Looking Back, Looking Forward: Re-Searching the Conditions for Curriculum Integration in the Middle Years of Schooling* (2007)

We have now set our destination and are ready to embark on our STEM journey. As in all journeys, we hope to make this trip meaningful and interesting. We capture meaningfulness in the level of rigor and relevance of the STEM learning experience and sustain interest through student engagement in the learning experience. The last two guideposts, R and E, emphasize the importance of addressing rigor, relevance, and engagement as a priority in student learning. In this chapter, we will examine the final two guideposts using a math-focused fifth-grade transdisciplinary unit titled *Feeding Fluffy* where students design a food container and delivery system to feed an exotic pet. Refer to the completed template in Appendix E for the details of each guidepost.

Feeding Fluffy: Summarizing the *W, H,* and *E*

The *Feeding Fluffy* unit was developed to focus on two disciplines—the core mathematics understanding of volume and the application of simple machine concepts. The Common Core State Standards for Mathematics list volume as one of the three critical focus areas for fifth graders to understand. The standards aim for a deeper conceptual understanding of volume, not just the idea that volume is the sum of the total of cubic units packed into the rectangular prism, but that the volume of rectangular prisms is additive and that the volume of known rectangular prisms can be used to find the volume of an unknown rectangular prism by stacking and adding.

In the *Feeding Fluffy* unit, more attention is devoted to solidifying the understanding of the additive aspect of this standard as students design an appropriately sized feeding container for an unknown animal.

The Essential Questions of this unit capture key understandings, emphasizing the mathematical understandings:

- What is the relationship between area and volume?
- How does the area of a rectangle help us find the volume of a rectangular prism?
- How is volume related to the size and shape of an object?

They in turn lead to the development of the Driving Question: "How can we use concepts of volume and simple machines to engineer a feeder for a given animal?" The Driving Question in this unit identifies an important secondary standard being addressed, a science standard about simple machines. The creation of a delivery system applies the concept of simple machines and therefore this standard is an important passenger in the STEM "car."

For this unit, we developed the following scenario (which will be fleshed out more in the second E Guidepost below):

> *The neighbors have suddenly gone away and they have left you a note asking you to feed their pet Fluffy while they are gone. They failed to leave a bowl to feed Fluffy and it is your task to design a bowl or container to accomplish this task.*

Before students begin designing a container that will hold the amount of food necessary for the animal, they will explore concepts of volume to fully understand how to apply the volume formula to rectangular prisms. Once these concepts are in place, the students will apply this volume knowledge and skill to design a blueprint with labeled dimensions for their containers. To complete the automatic portion of their feeder system, students will explore simple machines and how they help produce work. They will apply these concepts in developing a compound machine that would move their designed food containers to the animal. They must sketch a blueprint for this feeding system as well as provide evidence to justify their designs. The culmination of learning will be a presentation to an expert panel to share their design.

H: Considering the Sequence

Having identified the desired results, we begin thinking about the instructional approach that will lead to these understandings—the *H*. The particular milestones that form this guidepost represent the multiple standards that are part of this unit. Look at the completed template (Appendix E) and see how these interim stops along the route scaffold the learning toward the ultimate destination.

Teachers Often Ask

The question often arises, "If I am planning a transdisciplinary STEM unit, do I need to teach the disciplinary concepts to mastery before the students are asked to apply their knowledge and skills?"

It has been a prevalent idea in education that the progression from knowledge acquisition, to skill development, and finally to application is the best way for students to learn. We now know that application (problem solving) can also lead to knowledge acquisition and skill development. A problem or application scenario creates a need to know that focuses the students on seeking the knowledge and skills they will need to solve the problem. A good analogy might be using your brand new smartphone. Can you imagine having to read the owner's manual from cover to cover to fully understand how each feature works (content)? Only after this step can you begin practicing the use of each feature (procedures and skills). After completing that step, you finally can use the phone for your communication needs (application). No way! We start with what we need it for (application), then we learn the features we need by either doing or reading the manual (assimilation of knowledge), and finally we practice by doing over and over until it is second nature (skill development). The same occurs with learning in the classroom. The basics can be learned in highly rigorous ways within the context of a STEM unit. We will return to these decisions around whether to develop the skill first or allow it to develop intrinsically from the investigation later in this chapter in the R Guidepost section.

E: Planning for Evidence

After the desired results and content have been identified and set as a target, we describe the tools we will use to collect the evidence of learning. How will the students demonstrate their understanding of the concepts of volume? What evidence will they produce that can be used to evaluate their understanding of simple machines? For this unit, summative assessment, the students produce two detailed blueprints, one for their food container and another to show their delivery device. The formative assessments include performance-based activities where the students demonstrate their knowledge. The preassessments include making sure students can make linear measurements in both nonstandard and standard units, find area of rectangular shapes, and possess some ability in multiplicative thinking.

Now that we've introduced the unit, let's examine what it means to add rigor and relevance to the learning tasks.

R—Rigor and Relevance: How Will I Provide Opportunities That Will Lead to Increased Rigor and Relevance?

R: Defining Rigor and Relevance

Rigor refers to the level of thinking required by tasks, and *relevance* is how applicable the tasks are to the real world. The Rigor and Relevance Framework based on Willard Daggett's research and adapted from the International Center for Leadership in Education provides a useful way to think about these ideas (see Figure 5.1). Rigor is shown along the y-axis as a continuum of how we think, following Webb's Depth of Knowledge work. It begins at the lowest level where knowledge is simply acquired, and moves to the highest level where knowledge is assimilated. Daggett refers to this as the Thinking Continuum, or the Rigor Scale, which is used to determine how rigorous, or cognitively complex, a given task is, also called its level of cognitive demand (Stein et al. 2009). A problem that requires only memorization is at the low end of cognitive demand, whereas a task that requires students to make connections between and among ideas in new ways is a high cognitive demand task. Research has shown that using high cognitive demand tasks supports learning and leads to increases in

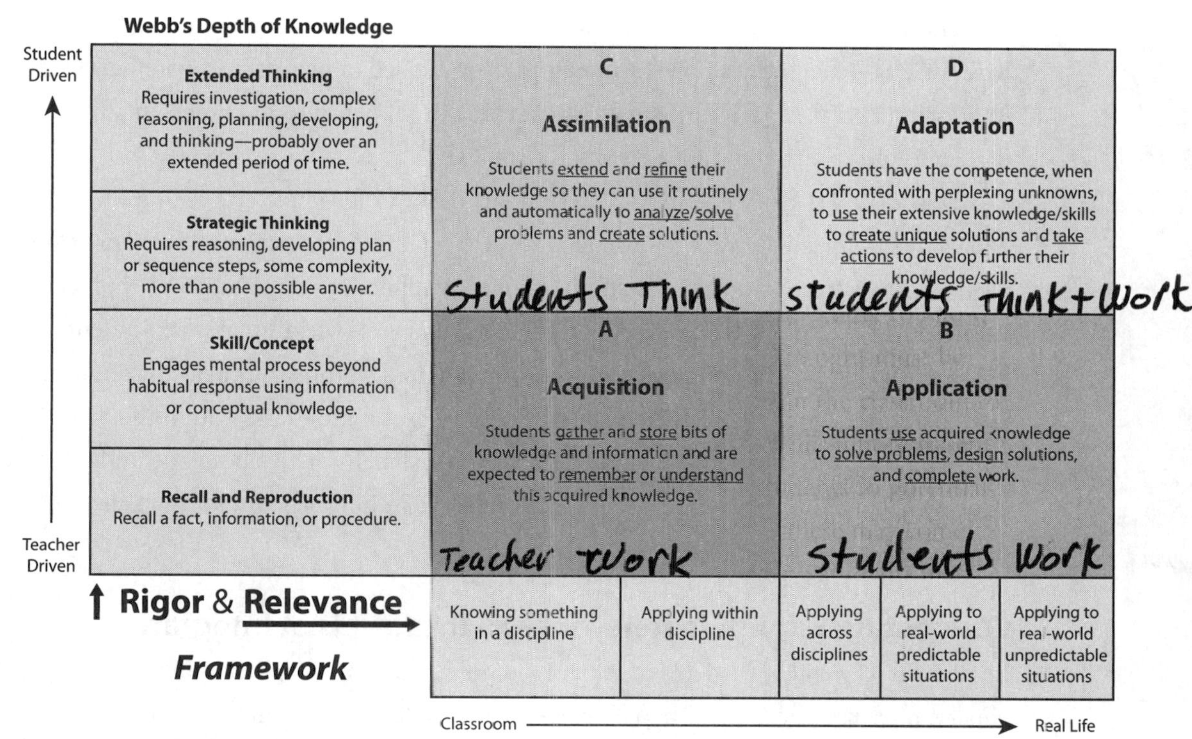

FIGURE 5.1 Rigor and Relevance Framework

student engagement. The scale also shows that rigor increases as students take more responsibility for the work, moving from teacher-driven tasks to more student-driven work.

Questions arise regarding whether increasing the rigor and cognitive demand at lower elementary grade levels is appropriate. Can rigor be achieved even though foundational skills are usually taught at the lower depth of knowledge range? This idea is addressed in the following excerpt from the book *How to Plan Rigorous Instruction*:

> *Rigorous thinking can be involved in learning even the most basic material. Students can learn the basics in highly rigorous ways. They can learn how to build adequate representations, organize those facts in some way, analyze and construct relationships among those facts, and make inferences beyond what is explicitly presented while they are mastering the basics. (Jackson 2011)*

The second dimension of the Framework, called the Application Continuum or the Relevance Scale, describes how knowledge can be applied in increasingly complex ways. Represented on the x-axis in the diagram, it shifts the application of knowledge from the world of the classroom to the world of real life.

Transdisciplinary STEM units help students move from simple acquisition of knowledge to assimilation of knowledge. These types of learning experiences also help move students from knowing something in one discipline, to integrating and applying knowledge across disciplines into new, unfamiliar, and real-world situations. Students are more likely to retain information if they are required to think more deeply about the ideas being presented. Through careful planning of exploration tasks within the STEM unit, a student required to think strategically and reason abstractly about the underlying concepts will absorb and retain this knowledge for a longer period of time and at a deeper conceptual level.

Connecting Rigor and Relevance to the Levels of Integration

In Chapter 1, we identified a hierarchy of integration levels: disciplinary, multidisciplinary, interdisciplinary, and transdisciplinary. These levels of integration correspond to increasing complexity and application of new knowledge as described by Daggett in the Rigor and Relevance Framework. In Figure 5.2, we overlay these levels of integration on the Rigor and Relevance diagram to show how they correlate to the framework. Learning activities done at the disciplinary level can move up within the acquisition and assimilation columns depending on the instructional approach. A science teacher, for example, may choose to use an instructional approach such as the 5E model to teach a concept. Even though the content is not integrated with other disciplines, the instructional approach can provide for inquiry-based, hands-on explorations and reasoning, which moves it up on the rigor scale. The same can be said for multidisciplinary integration. Because the disciplines are just related to a common theme and do not necessarily have a connection for understanding between them, a teacher may also choose to select an instructional approach that may give the students opportunities to use the knowledge or skills in open-ended investigations of their own choosing. Interdisciplinary and transdisciplinary STEM units in particular are naturally more rigorous and relevant because students are presented with complex problems that require them to apply their knowledge and skills to create solutions and take action in different and unfamiliar situations.

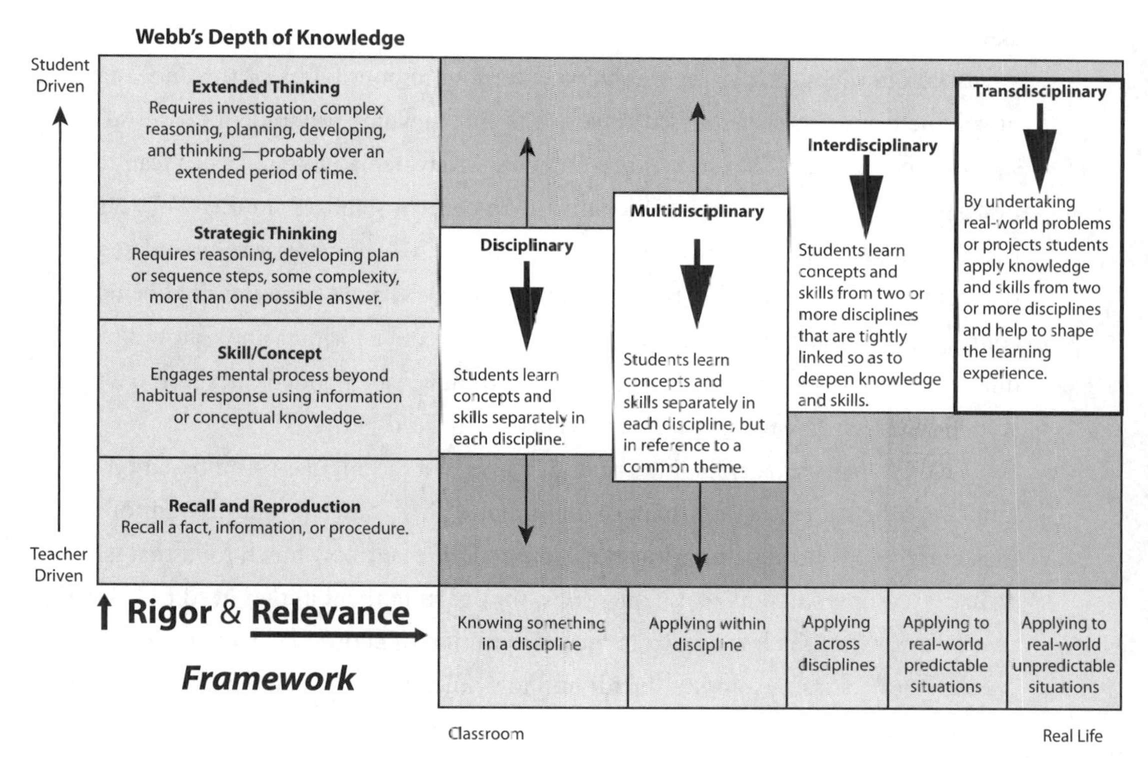

FIGURE 5.2 Rigor and Relevance Framework with Levels of STEM Integration Overlay

Rigor and Relevance in *Feeding Fluffy*

Now that we've looked at rigor and relevance in general, let's see how these ideas apply to the *Feeding Fluffy* unit. The following strategic guiding questions guide the "Rigor" and "Relevance" sections of the template.

Strategic Guiding Questions

- How will the learning experiences extend the student's thinking?
- What intriguing questions will foster greater inquiry to increase conceptual understanding?
- What opportunities will provide for increased relevance and real-world connections for the students?

Rigor

We asked ourselves, "How can we shift the learning responsibility for the understanding of volume from a teacher-focused approach to one in which the student takes ownership of the learning? What sort of rich and cognitively demanding tasks can lead to this understanding?" Even though we identified the content standards in the "W" section of the planning template, we now need to analyze how we can increase rigor and cognitive demand as students meet these content standards. The volume standard being addressed could be met by merely having students recall a formula and follow a procedure. The scenario, however, introduces an intriguing, open-ended question to engage student interest: "How big a container is needed to feed Fluffy?"

The task has students finding the dimensions of a rectangular prism based on volume—but the student has to think of the container first, based on the amount of food the animal needs. And before that, they have to figure out how much food that is! The scenario requires students to ask questions, thus putting them in that need to know setting that transfers the learning from mere acquisition of knowledge to assimilation.

As students start exploring the idea of how much food the container will hold and what the dimensions will be, other questions arise, whether posed by the students or prompted by the teacher. "Do different foods take up different amounts of space?" "How can we use standard units of measurement to find the volume of a rectangular prism?" "How do we stack standard cubic units to find the volume of a rectangular prism?" These strategic reasoning questions lead to student reasoning about additive volume, where students begin to see the connections between area, area measurement, and volume. Students then begin to connect the dimensions of the prism to the number of stacked cubic units, developing the formula for volume. Letting students reason first about the concept of taking up space rather than showing a formula for finding volume requires them to try to explain the underlying concepts that make the formula work. This reasoning before procedures sequence is a highly rigorous way of learning about volume.

After discussing their initial scenario of feeding Fluffy, the following twist is added to the task:

> *But the pet turns out to be a grizzly bear! How will this new information affect your design?*

The twist is added to guide students into making sure they are not assuming (most students think Fluffy is a cat or dog) and are asking clarifying questions. It also helps students begin thinking of constraints when designing the container. They must adjust the size, material, and maybe shape based on the new information (Figure 5.3).

RIGOR: How do I plan to meet this goal?

Identify tasks that can elevate student thinking, improve inquiry, and increase conceptual understanding.

The following tasks provide students an opportunity to extend their thinking.
Changing the scenario to include designing a food container for a grizzly bear.

- Students research how much food a grizzly bear eats.
- Calculate how much space that amount of food takes up.

Students are given the task of designing a food delivery system that can move the animal's food container using a variety of simple machines.

- Students must figure out which system of simple machines can work together to deliver the container.

Designing and presenting their final products forces students to apply the knowledge and skills from multiple disciplines and provide evidence to justify their designs.

FIGURE 5.3 "Rigor" Section for *Feeding Fluffy*

Relevance: What Opportunities Will Provide for Increased Relevance and Real-World Connections for the Students?

Providing inquiry-based tasks along the way may set students up for needing to know, but does it make it interesting enough for them to want to know? Thought must be given to ways to help students connect the learning that occurs within the classroom to the use of that knowledge beyond the classroom. "How can I make this unit represent the real-world condition?" "Is there an opportunity to introduce students to potential career opportunities that might be of interest to them?" Answers to these may come from student discussions or in thinking about the relevant information students may discover as they research the topic. For example, considering the scenario of feeding a dangerous pet animal, the topic of whether people should be allowed to keep exotic or dangerous animals as pets would naturally arise. Would a debate on the topic be appropriate? Could the students write a persuasive composition stating their position and validating it with facts? This is not crucial to the design of the animal feeder, but it brings relevance to the topic and is a great learning opportunity for students to research, speak, listen, engage in a Socratic seminar, debate, and integrate many of the English language arts content.

In the case of *Feeding Fluffy*, because the idea of feeding your next-door neighbor's grizzly bear seemed a little contrived, we decided to explore some more real-world applications for this same concept. By showing students a video of how some zoos feed the dangerous animals, and having a discussion on how these can be improved, the context becomes more authentic. The initial idea of designing a container evolved into designing an automatic feeding system to feed a dangerous animal. We now had a Driving Question that targeted the desired results and content and also made the why clearer for the students. Having students see real-world application of the way animals are fed moves the experience from a classroom activity to investigating a real-world problem (Figure 5.4).

Having ensured that the overall STEM unit is both rigorous in its expectation of student involvement and represents real-world issues, how will we entice the student into this experience? The final E Guidepost in the W.H.E.R.E. Model reminds us to consider how we will "hook" the students into the learning and keep them engaged.

RELEVANCE: Does the learning experience provide for relevant and real-world experiences?

Identify current topics and local issues that can make the tasks more engaging.

Students will take the role of engineers to design an animal feeder for dangerous animals.

- This brings relevance by connecting the learning to work done in the real world by engineers when they design solutions to problems.
- Discussions on raising animals in captivity and keeping dangerous animals as pets helps students vest themselves in the tasks and learning because these topics relate to real-world situations.

They will use the concepts of volume to design a container that holds the necessary amount of food.

- Any time you apply concepts and skills to solve a problem or fill a need, the relevance of the learning is at its highest level.
- The type of animal and what it eats provides relevance to the concepts of volume that the students are learning.

They will use the concepts of force, motion, and simple machines to make a machine that can lift, transport, and deliver the food to their designated animal.

Again, application of skills and concepts as they are being learned keeps the students connected to the learning and avoids it being disjointed from other concepts and skills.

FIGURE 5.4 "Relevance" Section for *Feeding Fluffy*

E—Excite, Engage, Explore: What Exploration Activities Will Excite the Students and Engage Them in the STEM Practices?

The STEM Lesson Guidepost Planning Template is not meant to be filled out from top to bottom. The components are part of a working document, filled out as ideas from one guidepost spark new thoughts in another section, where they interact or affect each other. This is very evident in the second "E" section of the template. This section further ties the content to the scenario developed in the "W" section.

The strategic guiding questions used to plan this component of the unit are listed below.

- What is the scenario that will hook the learner?
- How will the students be cognitively engaged?
- Which STEM practices will the students engage in?
- How will the explorations provide evidence of this engagement?
- How do the explorations address the Driving Question?

Excite: What Scenario Will Hook the Learner?

This section is where the scenario is further developed and refined. How do you create the "hook" activity that will quickly engage the students in the learning activities? It can be in the form of a discussion, a reading, a video clip, or a problem-based task. For the *Feeding Fluffy* unit, we adapted an existing engineering activity as our hook. As described above in the H Guidepost summary, the students are presented with this problem situation:

> *The neighbors have suddenly gone away and they have left you a note asking you to feed their pet Fluffy while they are gone. They failed to leave a bowl to feed Fluffy, and it is your task to design a bowl or container to accomplish this.*

After discussing their initial ideas the following twist is added to the task:

> *But the pet turns out to be a grizzly bear! How does this new information affect your design?*

The twist prompts the students to ask about the limitations or constraints that the solution must take into account, including size, weight, forces, and so on. This gets the students thinking about the concepts of volume and forces that are part of the learning goals. Students are now hooked and ready to learn about size and volume to help them design their containers (Figure 5.5).

EXCITE: What is the hook to excite the learner?

Create the scenario to engage the learner.

Students are presented with the scenario of a neighbor leaving town and they will need to feed the neighbor's pet, Fluffy, while they are gone, but the neighbors forgot to leave a container. Students will discuss constraints and other factors related to feeding Fluffy.

FIGURE 5.5 "Excite" Section for *Feeding Fluffy*

Engage—How Will the Students Be Cognitively Engaged?

Student engagement is broadly interpreted as being on task or participating in classroom activities, but this describes characteristics of just one domain of student engagement. The three domains of engagement are behavioral engagement, social engagement, and cognitive engagement (see Figure 5.6).

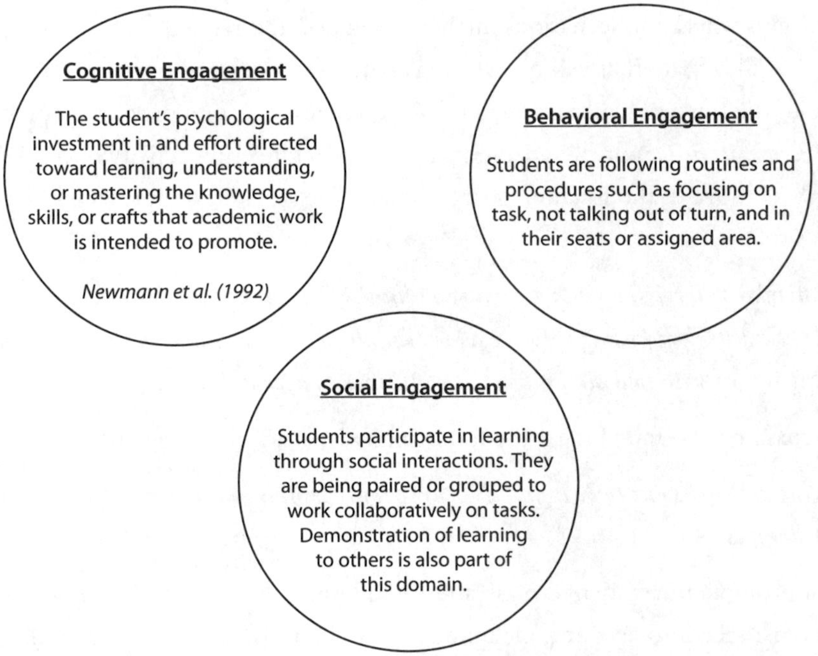

Cognitive Engagement

The student's psychological investment in and effort directed toward learning, understanding, or mastering the knowledge, skills, or crafts that academic work is intended to promote.

Newmann et al. (1992)

Behavioral Engagement

Students are following routines and procedures such as focusing on task, not talking out of turn, and in their seats or assigned area.

Social Engagement

Students participate in learning through social interactions. They are being paired or grouped to work collaboratively on tasks. Demonstration of learning to others is also part of this domain.

FIGURE 5.6 Domains of Student Engagement

Most teachers monitor behavioral and social engagement well. Asking them to keep students cognitively engaged, and monitoring that engagement, can be a more difficult task. Cognitive engagement in academic work has been defined by Marks (2000) as a "psychological process involving the attention, interest, investment, and effort students expend in the work of learning." Newmann et al. (1992) defined cognitive engagement in academic work as the "student's psychological investment in and effort directed toward learning, understanding, or mastering the knowledge, skills, or crafts that academic work is intended to promote."

Evidence of Engagement—Which STEM Practices Will the Students Engage In?

Keeping students cognitively vested in learning seems to be an undertaking that involves covert processes that are hard to monitor. Luckily, the STEM practices (see Chapter 1), and the practices of English language arts, involve overt behaviors that can be used as evidence to monitor both cognitive and social engagement.

In *Feeding Fluffy*, the mathematics driver has students reasoning abstractly and quantitatively about the measurements of volume and the size of the container necessary to feed Fluffy. They will *create and use models* to explore and convey their thinking. To understand the ultimate tasks, they must *ask questions and define the problem* they are trying to solve. Although most of the STEM practices could be applied to the learning in the unit, it is beneficial to focus on just three or four of the practices. This provides both you and your students with a much more targeted goal.

As we move to the final part of the E, exploration, we focus on including key practices, along with interactions, discussions, and questions that can help monitor them, as we select and refine exploration tasks (Figure 5.7).

ENGAGE: How will the students be cognitively engaged throughout the unit?

List the STEM practices that will be used as evidence.

Reason abstractly and quantitatively
Create and use models
Ask questions and define problems

FIGURE 5.7 "Engage" Section for *Feeding Fluffy*

Explore: How Will the Explorations Provide Evidence of This Engagement?

In this section, the exploration tasks that lead students to the desired learning outcomes tie together the other components of the planning template. The exploration activities should be thoughtfully conceived to support the Driving Question and answer the Essential Questions in the "Why" section, which in turn lead to deeper student understanding, increasing the level of rigor and cognitive demand. They should also engage students in disciplinary STEM practices that they will exhibit as proficient thinkers. In the *Feeding Fluffy* unit, the exploration activities focused on questions such as "What is volume?" For example, the Creating Prisms activity helps students develop a model of the concept of volume. Students draw a rectangle on centimeter grid paper of a given length and width. They then extend the shape to three dimensions by stacking centimeter cubes to a given height, creating a rectangular prism. To further connect the idea of volume to the volume formula, students engage in a task called Joel's Buildings, where they are given the dimensions of a building (rectangular prism) and are asked to build it using centimeter cubes. The idea is for students to find the total number of cubes they will need by multiplying the given dimensions. Students will now use this knowledge in designing their container for feeding Fluffy.

For the simple machines explorations, the students use a combination of simple machines to move a thirty-gram weight, five inches vertically off of the surface and twelve inches in any direction. This replicates the movement of their *Feeding Fluffy* container. The students use this knowledge of simple machines to develop a blueprint for the food delivery system.

Detailed descriptions for each of the explorations can be found in Appendix F. This detail reflects the actual lesson planning process. The information is organized to help make the lesson planning process more efficient and offer a consistent and structured model to follow (Figure 5.8).

The completed template for the *Feeding Fluffy* unit can be found in Appendix E.

EXPLORE: What activities will help students address the Driving Question?

List questions for students to investigate that will lead them to a deeper understanding of the content and skills.

Design a container to feed Fluffy: What is volume?
Creating prisms: How does the idea of area help to develop the idea of volume?
Joel's Buildings: How is volume related to the size and shape of an object?
Design a feeding system for the given dangerous animal: How is volume related to the size and shape of an object? How do simple machines help us do work? What is the relationship between force and motion?

Materials: centimeter cubes, grid paper, simple machine sets, poster paper, markers

Resources: zoo animal feeding video clip from YouTube

FIGURE 5.8 "Explore" Section for *Feeding Fluffy*

Think About It

This *Feeding Fluffy* unit is constructed using a transdisciplinary integration approach involving both mathematics and science concepts. The engineering context and problem scenario has students using concepts and skills from multiple disciplines to solve a real-world problem. The supporting exploration tasks are connected to the learning of those concepts and skills and represent rigorous and relevant experiences. In this unit's summative assessment, the students not only produce a product (two detailed blueprints—one for their food container and another to show their delivery device), they must also present their design ideas. In the next two chapters, you will be guided through your own use of this template to plan a unit of your own, first examining how a change of drivers impacts the route, and then creating a unit.

Speed Bumps

- What other relevant topics or scenarios could be used to address the same mathematics content driver?
- How might the hook activity be modified to appeal to a different group of students?
- What other type of exploration activities might you include to fit your own classroom needs?
- How could different STEM practices be incorporated into this STEM learning experience?

Changing Drivers

Who Is Driving the STEM Car?

The first step toward getting somewhere is to decide that you are not going to stay where you are.

—Unknown

As you know already, integrated STEM learning experiences provide for improved student cognition. But how can we think about these integrated STEM learning experiences through the lens of different disciplines? What would the unit be like if another content discipline were the driver? How would the experiences change to reflect the acquisition of different conceptual understandings? Would the entire unit need to be revised, or could we just make minor tweaks to the activities? Would the same Driving Question work or does that too need to be revised to match the new learning goals?

In this chapter you will examine the same STEM unit from Chapter 4 but with a new driver. The new driver will cause a shift among the other passengers, and some may end up being removed from the STEM "car." As you read this chapter, refer to the completed STEM Lesson Guidepost Planning Template for the *Wind Turbine* unit that is in the appendix. As we unpack our thinking around the impact that the new driver has on the unit, we suggest you make notes in your own working copy of the template. Record your ideas on how you might revise the unit based on your grade level, subject matter, and classroom experience, and add your own questions to contemplate in preparation for creating your own STEM unit.

In Chapter 4, the *Wind Turbine* unit was designed with the science standards as the driver and engineering design, mathematics, and English Language Arts (ELA) as the passengers. The unit had students develop an understanding of the relationship between changes in kinetic energy and the energy output of a system, using design changes in the wind turbine to generate their evidence.

Now, let's change the driver for this unit to the mathematics standards that were originally passengers and explore how this shift in intent will affect the unit.

W Guidepost—What and Why

Remember—the template is like a place mat for collecting ideas and information and does not need to be filled out corner to corner; however, it is critical to always begin with the *W*. *What* are the desired results, including big ideas, content standards, knowledge, and skills that you want the students to know and be able to do? And why are you teaching these concepts and *why* will the students care?

The first mathematics standard involves the major understanding that students must represent and analyze quantitative relationships between dependent and independent variables. The second requires real-world application of area, surface area, and volume for solving problems. How can the *Wind Turbine* unit address these conceptual understandings? Could students develop these knowledge and skills through working with the turbine? The investigation into the size and shape of the blades ideally lends itself to the application of these understandings. Deciding how changes to the size and shape of the blades impacts the energy output as measured by the electric meter gives students actual data that they can analyze. The development of a new Driving Question might begin something like this: "Do changes to the length of the blades or their surface area impact the amount of electrical energy generated?" From here, changes to the other guideposts might redefine the user experience and require a refinement to this early version of the Driving Question.

Now let's look at how the change in driver impacts the passengers. The engineering standards about iterative testing that lead to modifications in optimizing a design solution can easily remain as the main passenger. For the students to gather enough data to analyze, they will need to keep modifying the blades, using either changes to the length or the surface area. The science standard about energy transfer as a result of changing kinetic energy could still be used, but with its focus on argument from evidence, it is not tightly aligned to the intent of the mathematics standard and therefore it could be removed. But with the main student work now focused on the mathematics, the science passenger could still be used as an introduction to the kinetic energy concept, leading to richer experiences in another unit down the road, or if students have already been introduced to that concept, it could allow them to apply their understanding from earlier experiences in a new and different context. It may be an opportunity to provide follow-up work for students to review, practice, and apply their knowledge in a new way, allowing them to further solidify that concept in their mind. The same reasoning can be applied to the ELA standards: Do we keep them or replace them with something else? Will keeping them mean we need to change the scenario?

As you can see, making decisions about which passengers to add to the STEM car requires a holistic look at the entire set of grade-level standards from across multiple disciplines. In general, it's worth considering whether some other grade-level standards that are better aligned to the driver might be used, even though it might necessitate a whole new scenario. A new Driving Question that tightly links just math and science might take the unit from the transdisciplinary to the interdisciplinary level of integration. This is an acceptable option when adapting any unit so long as it addresses your goals.

Now that the new driver is identified, look at the completed Wind Turbine template and think of what new Essential Questions need to be added. Make a list of them on your working template. Think about how the new Essential Questions would be scaffolded so that they begin to create a route for the learning leading toward your new destination. As you work, are there any other standards that come to mind that could be added to further enrich the unit? As we examine the other guideposts, keep in mind how the Driving Question may need to evolve to capture the essence of any new information we incorporate into the STEM unit.

H Guidepost—How

Now that the driver is mathematics, the unit focus will be on generating and analyzing data. The intent of this guidepost is to think about the learning experiences that will enable students to come to a full understanding of this concept. How will you sequence these activities? In this example, you might ask yourself, "Will the students know the difference between dependent and independent variables?" If not, then how will you introduce it to them? Is there a lesson or activity that could be used to help them gain a working understanding of that relationship? Thought also needs to be given to the turbine construction process. Will the students know how to construct the wind turbine or should there be additional time allowed for this step? In thinking about the generation of the turbine data, you might ask yourself, "What are the constraints on the blades? Will my students have the foundational knowledge of the concept of surface area, and/or the basic skills for measuring that are necessary for the unit's success?" Should these skills be covered before introducing the unit or can they be incorporated into the unit's main investigations? Although having familiarity and facility with the skills ahead of time makes managing the project easier, separating that instruction from its relevant context makes it appear isolated and therefore the student may miss the need for it. How you choose to structure this instruction should be based upon your knowledge of your students and their needs.

Our answers to the previous questions help us to prioritize the sequence of individual learning milestones along the route. These are the main activities that will give the unit its core structure. For the new driver, the *Wind Turbine* unit milestones might look like this:

- creating an operational definition of dependent and independent variables
- learning how to measure surface area by making models
- understanding the wind turbine—how to build it, what the constraints are, the variables involved, and the criteria for success
- understanding the Driving Question: "Do changes to the lengths of the blades or their surface area impact the generation of electrical energy?" (Remember, this may be modified once we create the scenario to give the unit greater relevance to the student.)
- the report/presentation. (What product will demonstrate their understandings?)

If the goal of this unit is to recognize the relationship between dependent and independent variables, does this sequence of experiences help us to achieve that aim? Should other milestones be added? Will the process of the iterative testing of the blades generating changes in the electrical output of the system create sufficient and interesting data for the students to investigate? As you think of these questions, add notes to your working template. Would you change the order of the tasks? Add or modify a task? How would you unfold the learning to guide your students to the destination?

E Guidepost—Evidence and Evaluate

Now that we've begun to define the route, how will we know that the students have successfully reached the destination? For this, we need to think about what the students will produce to demonstrate their acquisition of these new understandings. How would students be able to demonstrate their understanding of the mathematical concepts that are now the driver in this unit? What kind of product would effectively communicate their knowledge of dependent and independent variables and of surface area? Could the product developed in Chapter 4 still be used as a basis for summative assessment even with the change of drivers? These questions frame what we want the students to produce as evidence of their learning.

The ELA writing standards used with the original driver could still be applicable with this new content. The students could produce a report in which they describe the input–output relationships of their turbine blade designs, relating it to the dependent/independent variables. The context for presenting information as described in the original scenario, however, is now different. The ability to analyze the numerical data information and present it in a clear and summarized manner is the intent of the new driver. The question then is, "What role could the students assume in the existing scenario that would make them more responsible for the analysis of the data than in just reporting it?" In a real-world situation, the engineering team would be responsible for gathering and analyzing blade performance data. Could the scenario change slightly so that it is now the engineering team who must gather and analyze the blade data to share with the management team of the Wind Turbine Company? At this point we formulate a general concept of the scenario and further details will be added in the second E Guidepost.

The ELA standards require students to "produce a clear and coherent writing in which the development, organization, and style are appropriate to task, purpose, and audience." Our changed role for students still works: the audience becomes the management team, and the purpose is to convey the analysis of the experimental data and identify which blade length and fin design are most suitable to the town's wind turbine location. The engineer could use the performance data to "introduce a claim and organize the reasons and evidence clearly" as part of the report, as called for in the ELA writing standard. In this summative assessment product, the students then must relate the changes in blade length or surface area to an increase in energy output, modeling the dependent and independent variables. This supports our decision to keep the science standards related to energy as a passenger.

As you ponder changes to this guidepost, think how you will assess the final product. Will you use a rubric? What specific criteria will be used and how will you distinguish the different levels across it? Does the rubric provide sufficient guidance to students so they will know what success looks like? Is that success attainable? Make notes in your working template and add additional questions now, as these will prompt your thinking when you later modify your own tools for assessing student understanding.

 REFINING THE DRIVING QUESTION

With the change in driver, we saw that the Driving Question needed revision too. At the beginning of this chapter we initially posed the Driving Question as "Do changes to the length of the blades or their surface area impact the generation of electrical energy?" Now that we have developed the summative assessment task with a focus on an engineering report of the blade performance data, we can revise the Driving Question by connecting it to a real-world context: "As the chief engineer of the Wind Turbine Company, what performance data can be used as evidence for the recommendation of an efficient turbine blade design?" This ties together the scenario and the product task, and identifies what the student should be investigating to provide a solution to the question.

Formative Assessment and Preassessment

Now consider how the formative assessment and preassessment in this unit will also need to change to reflect the new driver. The formative assessment section of the template asks us to think about the checkpoints we would want to implement to monitor the students' progress as they move toward the final destination. Will the formative assessment quiz developed for the original driver be an effective tool with this new one? If an introductory activity is used to develop an operational definition of dependent and independent variables, will the outcome of that experience be sufficient proof of their understanding? What other pieces of evidence does the school or district require as part of their grading system? As you think about these questions, make notes in your working template.

What core knowledge and applicable skills are necessary for students to successfully navigate the learning? How will you determine if these prerequisite skills are in place? The science driver required students to have an understanding of several key processes and an awareness of essential vocabulary. Are those necessary with this driver or are other factors more important? With the mathematics driver, it may be more important to make sure students have an ability to work with spreadsheets, being able to paste rules or formulas into cells to make the analysis of data easier. Do the students have the proficiency to graph the data or know how to gain access to programs that can translate spreadsheets into graphs? Knowing the difference between a dependent and independent variable is also important, but will a textbook definition be useful to students or would they benefit more from some type of hands-on activity that models that relationship? As we plan for the prerequisite needs for the unit, we may need to revise a task in the H Guidepost, or add an Essential Question in the "Why" section of the W Guidepost.

R Guidepost—Rigor and Relevance

Now we can begin to review the R Guidepost to see how we can add greater rigor and depth to the overall learning experience. Looking back at the Rigor and Relevance Framework described in Chapter 5, we see one way to increase the rigor in the unit is to shift the burden of learning from the teacher to the students. Making students more cognitively engaged in the learning task increases their ability to initiate discussion, propose new ideas, and take ownership of the experience. In this example, we need to ask ourselves, "Would the mathematics driver require modification of the existing 'Rigor' section or does the focus on the variable relationship remain the same, whether it is tied to the science or mathematics content?" Although the science rigor included a focus on the kinetic energy concept, the mathematics rigor could skip that and just focus on the parts that relate to dependent/independent variable pairs. In other words, students could still investigate changes to the blade design and use that as part of their data analysis report. They could also look to combine different factors together to see how they influence each other. In either case, allowing the students to take ownership of the parameters they choose to investigate moves them into assuming a more active role in the unit, thus making it a more rigorous task.

We should ask ourselves, "What opportunities are there to connect the work of data analysis to real-world situations?" How could the existing ideas in the "Relevance" section of the template be refined to be more representative of the intent of the mathematics driver? The "Relevance" section asks us to consider what connections to the community, whether local or global, or connections to self would be of interest to the students. This section is constantly evolving because it should incorporate timely world events and local issues. Even a STEM unit created last year might benefit from a revised scenario if new local, national, or world events make the topic more timely and of greater interest. Using your own grade-level content expertise and classroom experience, make a list in your working template of the topics that interest your students. If you are unsure, let the students generate a list of topics they find interesting. This list can serve as a basis for weaving relevance into your learning experiences and should be updated frequently.

E Guidepost—Excite, Engage, Explore

Finally, we approach the last E Guidepost. How will the students be engaged, both cognitively and physically, in the learning experience? This section of the template provides an opportunity to refine and further define the scenario to be both engaging and interesting for the students.

How can you refine the hook to capture their interest and lead them into the main focus of the unit? Just as we've seen in the other guideposts, the change to the mathematics driver has created the need to modify the original unit template. What changes would you make to the scenario now that would encapsulate the revisions we've made so far with the new driver in the other guideposts? If we use the student product as described in the summative assessment section, then the scenario might go something like this:

> *The Wind Turbine Company has received a request for proposal for an efficient wind turbine design based on research. As the chief engineer for the company, you have been tasked with determining which blade length and fin surface area would be best suited for the project. You must present your findings and recommendations to the Wind Turbine Company management team.*

Can you see how the addition of details makes the scenario more credible and engaging for the students? Are there other scenarios to hook the students? Make a list of possible scenarios in your working template. Do these different scenarios have things in common? Do they all include some form of investigational procedure and data analysis as part of the work? Do they include some method of presenting the findings in a summary report? Do they include a role for the student to assume in the work? The creation of the scenario sets a context for the learning. It can also uncover the kinds of cognitive tasks that the students should be involved in, leading to the identification of appropriate STEM practices.

Choosing STEM Practices

Although most of the STEM practices might be applicable, which ones should you focus on? Narrowing down the list of STEM practices to just three or four provides a better opportunity to offer strategic and targeted instruction and therefore greater attention being given by the student. Which mathematical practice standards would you select for this unit? What behaviors would you expect the students to exhibit if they were attending to this practice? Use the template to list the STEM practices and the behaviors you want to see the students emulate. These notes can help you later as you prepare a rubric to evaluate the overall student performance.

Reflecting back on the Essential Questions posed in the W Guidepost and the milestones proposed in the H Guidepost, we can now focus our attention to adding details to these individual explorations. Ask yourself, "How will the students explore the relationship of dependent and independent variables?" Is there an activity that could either help introduce the concept to them or provide experiences around which they could develop their own operational definition of these terms? Ask yourself similar questions about the other milestones and activities, and collect those details in the "Explore" section of the template. As described in Chapter 5, there is no one approach to the details you compile here. Add the information that you need to further plan as you build your lessons. It is also good to consider any constraints due to time, materials, or students' previous experience: How can we balance the need to increase rigor by allowing the students to take ownership of the investigation, yet at the same time manage with the quantity of materials available? Can we group together some of the Essential Questions so as to shorten the time needed to complete the unit? These questions allow you to begin the tactical planning of the unit, making it conform to the realities of the classroom.

Review the original *Wind Turbine* template and your own working template and look over the changes we made for accommodating the mathematics standards as the new driver. Did we make all the necessary adjustments to meet the expectations they required? Is the Driving Question still supported by the Essential Questions? Does the scenario make sense and is it applicable to the real world? Is the student product of sufficient rigor to convey understanding of the driving standard?

Think About It

As you see, the decisions you make along the way influence the direction of the STEM journey. The selection of the driver impacts all of the sections of the template. As you become more familiar with the STEM lesson guideposts, you will find it easier to craft your own STEM units. Now that you have seen how the change of driver impacts your planning, you will have the chance to revise one of the other example units in the next chapter.

Speed Bumps

- What other content could you have added as the driver? How would that decision impact the development of this template?
- Do you start with the driver and create a scenario? Or can the scenario determine what content will be selected? Does it matter which comes first?
- How can you collaborate with others to develop an integrated approach within your classroom?

Now It's Your Turn

You Drive the Car

It is our choices that show what we truly are, far more than our abilities.

—J. K. Rowling

Looking back at the STEM units developed as examples so far in this book, we have seen different drivers leading the journey. In Chapter 3, the engineering design process was the driver for *Engineering a Hat*. In Chapter 5, the *Feeding Fluffy* unit used mathematics as the driver. What would each of these units look like if the driver was changed to science? How would that change impact the information in the different lesson guideposts, the unit's Driving Question, or the final student product? Do you think that the "Rigor" and "Relevance" sections would need modification or could they be kept as they are? Is there a need to select other STEM practices as the focus for cognitive engagement because the content standard is different? Consider these questions as you begin the process of revising a completed STEM unit template.

Your Task

Consider how you would revise either *Engineering a Hat* or *Feeding Fluffy* from the perspective of your science standards. Use your own grade-level science standards to choose a new content driver. Create a new STEM Lesson Guidepost Planning Template using the original in the appendix as your guide. The set of questions that follow can help guide your thinking about how the template could be revised to meet the demands that this new content driver brings to the unit.

What and Why

- List the science standards you wish to cover in the STEM unit.
- Think about the desired results from adding these new standards.
- What changes are required to the Essential Questions?
- How would those changes impact the Driving Question and affect the scenario?
- What kind of scenario would reflect the intent of these science standards?

How

- What major learning experiences will develop student understanding?
- Is the sequence of these experiences logical and structured so as to promote conceptual development leading to understanding?
- Do the learning experiences support the Driving Question and provide context for the Essential Questions?

Evaluation

- What student product will be sufficient proof of students' learning and demonstrate conceptual understanding?
- What measures will you use to monitor student progress toward the unit's overall goal?
- Did the change in the driver necessitate a change in prerequisite skills and knowledge?

Rigor and Relevance

- Do the learning experiences extend student thinking beyond basic knowledge?
- Are the experiences presented in a way so as to put the burden of learning on the students, making them active participants?

- Can the scenario be improved so that it builds on current topics of local, national, or global interest?
- Will the students see themselves as possible participants in a related career?

Engagement

- Will the hook work with the new driver or does it need to be modified to fit the new learning goals?
- Which STEM practices would relate to the work of the unit and model real-world applications?
- How will the explorations provide the necessary experiences to address the essential questions?
- What materials, planning, or structure might be needed to manage these explorations?
- How do the explorations support the goal of the Driving Question?

Look Back

After you change the unit template, it is important to look at the template as a whole and make sure that revisions to later sections didn't unintentionally affect earlier ones. Check the sections to make sure the following conditions are still met:

- The standards are sufficiently covered.
- The Driving Question will capture student interest.
- The evaluation tools address the intended learning goals.
- The student product reflects the multiple standards (the driver and the passengers).
- If you added new explorations, then you revised the Essential Questions as needed.

Starting from the Beginning

In the rest of this chapter, you will be completing your own STEM Lesson Guidepost Planning Template for a STEM unit called the *Art Pedestal Project.* Your task is to use the *Art Pedestal Project* unit as a foundation for completing your own STEM Lesson Guidepost Planning Template for the grade level and content driver you wish to target. This learning experience will facilitate your own planning of a STEM unit.

This example is based on a scenario that was originally designed for intermediate and middle school students, centered around a set of social studies learning experiences in which students develop an understanding of the importance of working together within their community. The C3 Framework for Social Studies State Standards suggests that by the end of eighth grade, students have some experience with the concepts of civic virtues and democratic principles. In other words, it calls for an awareness of "the virtues such as honesty, mutual respect, cooperation and attentiveness to multiple perspectives that citizens should use when they interact with each other on public matters."

Students are asked to create a pedestal using straws, tape, and paper. They must first create a scaled drawing (blueprint) of their design and get it approved prior to "construction." The surprise or "aha" for the students in the original *Art Pedestal Project* is when the great pieces of art are revealed: small stuffed animals, or Beanie Babies. These items work nicely as they are lightweight and flexible enough for the students to mold to sit upon the pedestal they build.

Ah, perhaps you might be thinking, social studies and STEM? How do those fit together? Well, another goal of the following experience is to demonstrate how disciplines can move into the driver's seat, while the science, mathematics, technology, and engineering disciplines become important and influential passengers on the journey. Certainly students have either seen in person or in pictures a variety of statues that are a part of their community. How do these symbols reflect the values of the community? What is the decision-making process that occurs to get the statue erected? What factors must be considered in the design of the pedestal, including the weight of the object, the weather and environment where it will be displayed, and the cost of different materials?

Art Pedestal Project STEM Unit Overview

In this unit, the goal is for students to work together to create a pedestal that is both aesthetically pleasing and cost-effective for the town, which has received a donated statue. The students must work together to plan, design, and build a pedestal within budget. The main goal is to highlight the virtues all citizens should exhibit when they work together to solve a problem. How does a community decide what the statue will look like, where it should be placed, and how it should be displayed? Think about each of the guideposts and how you would complete that section based upon your own content decisions.

Developing the Template for Your Own Unit

We realize that there are many different ways to approach this unit. The state standards for the different disciplines and their grade-level placement are many and varied. We encourage you to use your own grade level as a familiar starting place in thinking about how you would complete the template. We have provided one possible completed template in the appendix as a guide to your own template development.

Let's begin by introducing our scenario and Driving Question for this unit. This is just one possible destination; you may decide to change it. That is up to you. You must decide what the student product will be, which standards the unit will address (both as the driver and the passengers), what route you will take, and how you might ensure sufficient engagement, rigor, and relevance. In other words, you are creating your own STEM car.

Scenario

The town has just received word that a citizen is going to donate a beautiful piece of art to the community. The town will need to have a pedestal built to support this art piece. The Board of Selectmen for the town council have decided to hold a competition for the community for the design and construction of this pedestal.

THE DRIVING QUESTION

How can our team build a structure to support this work of art, constructed within budget and finished on time?

Planning the *W*—What and Why

As stated earlier, even though the template does not have to be filled in from corner to corner, you will always want to begin with the W Guidepost. How will you determine the driver and the passengers for this unit? *What* are the desired results, including the big ideas, content standards, knowledge, and skills that you want the students to know and be able to do? *Why* would the students care about knowing and understanding this content? What Driving Question will you create to fit the driver and passengers in your STEM car? (We provide one above, but you might choose something very different!)

Which discipline standard have you decided is the driver? Which ones will you choose as passengers? Did you choose science as the main standard? Will the focus of the student work be on the concept of structure and function looking at the structure itself? Or perhaps it will be about investigating the properties of different materials used for construction? Maybe you choose mathematics as the driver. Is your purpose for the unit to apply concepts of ratio and scale models, relating the design blueprint to the actual construction? Maybe it is to apply and practice problem solving using computation of decimals and fractions as in a budget summary sheet. Yet maybe you preferred social studies to be the leader for this STEM journey. Will the focus be on learning about how local government works, the governmental decision-making process? Are there any English language arts standards that you might wish to incorporate? Does everyone in the community want the statue? Could students write a persuasive letter arguing either for or against the pedestal design? All of the above might be ways to identify the driver and the passengers.

What Essential Questions have you developed to frame the learning experience? Did you modify the Driving Question to better suit your intended destination? Remember to consider the following:

- List the standards you wish to cover in the STEM unit.
- Think about the desired results for these standards.
- What are the Essential Questions?
- How would those changes impact the Driving Question and affect the scenario?
- What kind of scenario would reflect the intent of these standards?

Planning the *H*—How

Having selected the driver and the passengers, created the Essential Questions, and possibly modified the Driving Question, you have already started thinking about how to move students toward the destination. Will the students build a model art pedestal? Will the students need to have a prerequisite set of skills in place to be successful? If they are building the pedestal, will there be limitations, such as time or materials? What major tasks should the students achieve that will move them toward the goal? Are these tasks sufficiently rich so as to inform your teaching and allow for opportunities to monitor and assess student progress? Remember to consider the following:

- What are the major learning experiences that will develop student understanding?
- Is the sequence of these experiences logical? Structured in a way as to promote conceptual understanding?
- Do the learning experiences support the Driving Question and provide context for the Essential Questions?

Considering the *E*—Evaluation and Evidence

In the original STEM unit, the development of a blueprint for the design of the art pedestal provided an opportunity for assessing student teamwork and collaboration. This became part of the individual and team evaluation. The students also made a presentation to the town explaining the process they used to design and build the pedestal. Another important piece in the original unit was the use of a budget summary sheet. The budget summary sheet provided students with some constraints to their pedestal construction. It required the team to carefully preplan their structure's design and to make group choices before and during the construction process. See Figure 7.1.

Materials & Labor Cost	$100.00 Building Budget	Amount of Materials Used	Total
Straws	$2.00 each		
Scissor rental	$4.00 each		
Tape	75 cents per 3 centimeters		
Paper	$1.50 per sheet		
Labor cost	$3.00 per 10 minutes per person Number of employees Teams of 5 or more get a rate reduction to $2.50 per 10 minutes		
Total amount spent			

FIGURE 7.1 Budget Summary Sheet

As you design your unit, remember to think about the following: Will the product effectively communicate the students' acquisition of the knowledge and skills required for the driver and passengers in your STEM car? Will it be an individual or a collective team product? Have you given thought to the interim formative assessments that you will use to monitor student progress toward the unit's goals? Have you considered how you would use them to guide and inform your instruction? The selection of the driver and passengers may require particular skills or knowledge. Have you considered what those are? Will the students need a prior introduction or can these skills be introduced as part of the unit's ongoing investigations? As you begin formulating your assessment plan, remember to revisit your Driving Question and revise it to reflect your unit's product. Remember to consider the following:

- Will the students' product be sufficient proof of their learning and demonstrate acquisition of their conceptual understanding?
- What measures will be employed to monitor student progress toward the unit's overall goal?
- Did the change in the driver necessitate a change in the prerequisite skills and knowledge that are necessary for student success?

Improving with the *R*—Rigor and Relevance

The original unit requires students to preplan their structures, with limited tools, materials, budget, and time. They must make choices and decide what materials, procedures, and tools they will use. This makes them fully responsible for their work—one way to increase the rigor of the work. To make the unit more relevant and tie it more closely to the social studies curriculum, the students take pictures of statues in their community. A local historian also provides insight into the history of the structures and the issues they raised at the time of their erection in the community.

In your version of the *Art Pedestal Project*, how will you shift the burden of learning to the student? What techniques or strategies are you going to use? Is your student product sufficiently rigorous? Does it make the students responsible for assuming the heavy lifting in developing an understanding of your driver? What have you done to make the work more interesting or enticing to the student? What topic of local or global interest would the students be motivated to learn more about? Does it offer the students the opportunity to explore potential careers? As you consider the R Guidepost, check back to make sure that the major milestones are well marked, that the Essential Questions target the learning goals, and that the scenario captures the core understanding of the driver. Remember to consider the following:

- Do the learning experiences extend student thinking beyond basic knowledge?
- Are those experiences presented in a way so as to put the burden of learning on the students, making them active participants?
- Can the scenario be improved so that it builds on current topics of local, national, or global interest?
- Will the students see themselves as probable participants in this as a career or a professional choice?

Motivating with the *E*—Engagement

The original unit's scenario hooks the students and provides a role for them in the final activity. The work includes several of the STEM practices: asking questions, constructing a viable argument, and applying mathematical principles. In the "Explore" section, more details and ideas are added to the different activities that will move the learning forward and are tied back to the Essential Questions developed in the W Guidepost. As these details emerged, we revised some of the prior work in the other sections. For example, adding the opportunity to take pictures of local statues and look at famous statues from around the world to better understand what different pedestals look like led to the addition of the Essential Question, "What observations need to be made and what questions need to be answered before creating the design for the pedestal?" This, in turn, sets up the "need to know" that engages students.

What hook have you created for your unit? Does it provide a role for the students to assume? How does it connect to the Driving Question and does it support the unit's culminating project? Will the existing STEM practices remain or are other practices better suited to the demands of your driver and passengers? Can new activities be added or existing ones be revised to encourage greater student interest? Remember to consider the following:

- Will the "hook" work with the new driver or does it need to be modified to fit the new learning goals?
- Which STEM practices would be appropriate to the work of the unit and model real-world applications?

- How will the explorations provide the necessary experiences to address the essential questions?
- What materials, planning, or structure might I need to manage these explorations?
- How do the explorations support the goal of the Driving Question?

Throughout the course of this book, we have laid out the *how* of developing your own STEM unit. As classroom educators, you can now begin to use your standards and curriculum to develop, enhance, and personalize your own STEM instruction. In our work with districts and teachers from across the United States and around the world, we found that most individuals in this process of change go through a predictable cycle and that the change is most often successfully implemented when all of the participants can share in the vision and aspirations of what that change will do for them. It is important to recognize that this process of transformation takes time and perseverance and requires the dedication and support of all the stakeholders within the system—administrators, teachers, parents, and the community. But what does this change look like in the context of a learning community setting, either within an individual school or across an entire district? In the next two chapters we share two different stories, one from an individual school on members' use of the STEM Lesson Guideposts to initiate the development of their own units, and the other from a county district, describing their members' transformation process toward greater STEM integration. Let us continue our STEM journey.

Think About It

We hope that you have experienced the template in a way that will help you and your team begin to develop your own STEM units. We do know, and research shows, that for students, integrated STEM experiences engage them in activities that involve the application of knowledge and skills, and this active engagement in learning solidifies their cognition. The social and cultural factors are also fundamental to all learning experiences and are particularly important in integrated experiences, where typically students are required to work with each other, actively engage in discussion, participate in joint decision making, and collaborate in problem solving.

Speed Bumps

- How did the inclusion of the social studies content help you think differently about developing STEM units?
- What influenced your decision to include other content passengers in the STEM car?
- How did the inclusion of the budget requirements expand the STEM focus of the unit?
- Are there other ways to incorporate different STEM disciplines in the *Art Pedestal Project*?

Changing the System

One School's STEM Journey

Although a major focus of change initiatives is on the individual changing, professional development can succeed only with a simultaneous attention to changing the system within which teachers and educators work.

—Loucks-Horsley et al., *Designing Professional Development for Teachers of Science and Mathematics*

Implementing a STEM program, like any new approach, is about changing the system. We know, and research has pointed out, that there have to be certain key elements in place to enable success:

- first and foremost, a supportive administration that will help provide the time for teacher collaboration and professional development
- access to the resources needed for curriculum and supplies
- a good system of communication to the parents describing what STEM learning is and how it will affect their children
- the establishment of community partnerships to help support the teachers and perhaps provide needed resources and expertise.

Although not intended to be a complete list, these factors provide the foundation for schools that want to implement STEM teaching and learning.

The Concerns-Based Adoption Model

Implementing change within a school or district involves many issues: time, resources, strategic planning, curricular changes, mentoring/coaching, professional development, communication, and most of all, buy-in from all involved. In other words, you have to bring about change within the entire system for any initiative to provide the outcomes you are expecting.

Susan Loucks-Horsley and colleagues (2010) describe one model for change:

> *One model for change in individuals, the Concerns-Based Adoption Model (CBAM), applies to anyone experiencing change, that is, policy makers, teachers, parents, students. The model (and other developmental models of its type) holds that people considering and experiencing change evolve in the kinds of questions they ask and in their use of whatever the change is.*

The model examines how people experience the process of change through the types of questions they ask. In general, the early questions are more self-oriented: "What is this?" "How will it affect me?" "Why are they doing this to me?" The worry is focused on the individual and the discomfort change can bring. When these questions are resolved, new questions emerge that are more task-oriented: "How do I do it?" "How can I use these strategies effectively in my classroom?" "How can I organize myself to implement this new way of teaching?" The concern shifts from the effect on the individual to the impact on the process or the disruption of classroom routines. As the change becomes more accepted and familiar, the questions focus on the benefits of the change: "How could I use this in other ways?" "How can I share with my colleagues how this works for my students?" "How are others using this approach?" The concern now is more altruistic, being able to share the newfound knowledge with others and to make it better.

The chart (Figure 8.1) describes the stages and typical expressions of concern that implementing anything new can bring about.

CBAM highlights the way people respond to change. Being aware of these internal concerns and listening for the kinds of questions being raised when discussing a new instructional approach can help you head off any potential disruptions and mitigate

Stages of Concern	Expressions of Concern
6. Refocusing	How can I expand and implement this with other areas of my teaching? Exploring more benefits.
5. Collaboration	How can I relate what I am doing to what others are doing? Can I cooperate with others?
4. Consequence	How will it affect my students? How can I refine it to have more impact?
3. Management	I seem to be spending all my time getting materials ready. How can I do this more easily?
2. Personal	How will using it affect me? I'm uncertain, unclear, and unsure.
1. Informational	I would like to know more about it. Gaining awareness of the new initiative.
0. Awareness	I am unaware or not concerned about the new initiative.

FIGURE 8.1 CBAM Model Stages of Concern (adapted from *Taking Charge of Change* by Shirley M. Hord, William L. Rutherford, Leslie Huling-Austin, and Gene E. Hall. Copyright © 1987 by ASCD. Published by ASCD. Reproduced with permission.)

challenges. As you read the rest of this chapter, think about your own experiences with trying something new. Can you identify your own stage of concern as you begin the process of implementing STEM into your curriculum?

In the following section, we share an example of applying the STEM Lesson Guidepost Planning Template to the curriculum planning process. The Killip team members share their own challenges and successes as they moved ahead to bring about major instructional change in its school. This is the story of their journey as shared by Ted Komada, the Killip School STEM Teacher, and Joe Guiterrez, Killip School Principal.

One School's STEM Journey—W. F. Killip Elementary, Flagstaff, Arizona

W. F. Killip Elementary School has taken these suggested implementation and change strategies to heart and ratcheted up the natural connections between their own standards and the twenty-first-century skills. The teachers and students really blossomed and came together to create an effective implementation of a STEM approach.

Overview of Killip Elementary School

In a school that draws nearly all of its 500 students from within a one-mile radius, hosts a 43 percent inward/outward mobility rate, and sees 98 percent of its students on the free lunch program, "going to school" means far more than learning your letters and numbers. Most of the Killip parents work multiple jobs. They do not have the financial means to pay for their children to participate in outside activities and even if they did, many do not have the transportation accessible to get to these nonschool experiences. Research indicates that these types of nonschool experiences, beginning at an early age, provide students with a variety of background knowledge and conceptual understandings about what exists in the world, how it works, and what their role is as they interact with it. Children living in poverty don't have access to these experiences and as a result, a debilitating experiential gap begins to form at a very early age.

At Killip, we strive to supplement the experiences our students do not get at home. We believe the answer is teaching reading, math, and writing through STEM content and experiences. Students can learn their foundational academics while closing that experiential gap as they develop solutions to real-world problems that are highly engaging, motivating, and relevant. This is the curricular foundation for *all* children at Killip.

A Three-Year Plan

Over the course of three years, we made the systemic and procedural changes necessary to develop units of instruction that were based on STEM content, but allowed for teachers to deliver instruction for all content areas.

Year One

In year one of the project, we focused early professional development efforts on gaining a conceptual understanding of what STEM was and how we could begin to plan our instruction for it. We then sat down with our teachers, instructional coaches, and STEM coordinator, one grade level at a time, and selected the Next Generation Science Standards we thought we could integrate with our reading and math standards. Many of our units in year one were very heavy on science investigations. First, students would read about science phenomena to build background knowledge, and these texts were used to address the reading standards. Then they would engage in some hands-on science activities to investigate the phenomena. Any time the activities involved using numbers and data, we would take advantage of the opportunity to engage with our mathematics standards. We arrived at the end of year one proud and amazed at schoolwide changes that had been made in such a short amount of time. The teachers noticed that students were more engaged in their learning and that their interest in the science topics motivated their attempts at learning to read.

Year Two

In year two of the project, we realized that although hands-on investigations provided students with the opportunity to gain content knowledge, they frequently did not provide the opportunities to apply that knowledge through a different context. We felt that many of our units provided a fragmented learning experience with students moving from text to text, investigation to investigation, activity to activity without any overarching goal or purpose. To bring some cohesion to our units, we began looking at the distinction between a series of instructional activities related to a theme and a set of activities related to creating a culminating project or to solving a problem. This targeted end-of-unit goal gave us something to connect all of our instructional activities together. As we began our planning, we realized we were collecting many more experiences than were realistic to implement. The scope of the unit had begun to creep beyond our initial expectation. We then asked ourselves, "Does this instructional activity help develop an understanding or skill that will allow students to create that final project or solve the problem?" If yes, we chose to keep the activity. If not, although it might have been a fun activity, it had to be discarded because it was taking up instructional time that could be better aligned with the goals of the unit.

Throughout year two, we engaged in professional development activities that built a foundational understanding for our staff. We continued to engage with whole-staff Problem-based/Project-based Learning (PBL) professional development including in-class coaching and mentoring as we continued to develop and revise our STEM units. Aligning all instructional activities in the unit to one end-goal allowed us to take a major step forward in the development of our STEM education model.

Year Three

In year three of the project, our focus shifted to evaluating the type of problems we framed our units around. We worked with the concepts of real-world problems and problems that were personally or locally relevant to our student's lives. Reframing our units to focus on solving these kinds of problems helped to increase our students' engagement and motivation for learning. Students became more impassioned and engaged in their learning because it was on a topic or issue that mattered to them. They could see and make meaning of this new content knowledge in the relevant context. The students themselves saw the value and were driven to identify or create solutions to the problems that existed in their lives or in the lives of those around them.

Challenges

As you might expect, we have encountered challenges as we have implemented this new form of instruction. One of these was the consideration of a more structured approach to solving problems. As we continued our STEM unit development with projects and problems that the students were trying to solve, we found that the students often lacked a well-developed plan or approach to the problem. By engaging our teachers in PD focused on the engineering design process and how to explicitly teach it, we can provide our students with a more systematic approach to developing better problem-solving skills. This focus will provide students with a more succinct, organized, and successful approach to implementing problem-solving strategies while ultimately freeing up instructional time for teachers.

Another area of difficulty we encountered with the implementation of project-based learning was how to get students to work effectively together. When the tasks are simple and require low levels of creativity and critical thinking, communication and collaboration among students is fluid. Yet as the problems and tasks became more complex, we found that communication and collaboration among students began to break down and progress toward the objective came to a sometimes fiery halt. We recognized that we must be more explicit in our teaching of the twenty-first-century skills of communication and collaboration.

A Sample Unit from Killip

The following STEM Lesson Guidepost Planning Template highlights the ideas of planning that went into one of the Killip Elementary Third Grade STEM units. This Rover STEM unit focused on the science concepts of force and motion as its driver and included a full load of passengers—English language arts (both reading and writing), social studies, and mathematics. In the unit, students were to design the wheels for a rover that could be used on any type of Martian surface. They had to plan and conduct their investigation, gathering measurement data to support their claim as to which wheel design was most effective.

W—What Needs to Be Learned and Why?

WHAT are the desired results, including big ideas, content standards, knowledge, and skills?

List the content standards and what the students will know and be able to do.

MAIN STANDARD (Driver)

Science

3-PS2-1: Plan and conduct an investigation to provide evidence of the effects of balanced and unbalanced forces on the motion of an object.

3-PS2-2: Make observations and/or measurements of an object's motion to provide evidence that a pattern can be used to predict future motion.

SECONDARY STANDARDS (Passengers)

Reading

3.RI.1: Ask and answer questions to demonstrate understanding of a text, referring explicitly to the text as the basis for the answers.

3.RI.3: Describe the relationship between a series of historical events, scientific ideas or concepts, or steps in technical procedures in a text, using language that pertains to time, sequence, and cause/effect.

3.RI.4: Determine the meaning of general academic and domain-specific words and phrases in a text relevant to a grade 3 topic or subject area.

3.RI.8: Describe the logical connection between particular sentences and paragraphs in a text (e.g., comparison, cause/effect, first/second/third in a sequence).

Writing

3.W.2: Write informative/explanatory texts to examine a topic and convey ideas and information clearly.

3.W.8: Recall information from experiences or gather information from print and digital sources; take brief notes on sources and sort evidence into provided categories.

3.W.10: Write routinely over extended time frames (time for research, reflection, and revision) and shorter time frames (a single sitting or a day or two) for a range of discipline-specific tasks, purposes, and audiences.

Social Studies

3-1.1.PO 1: Use timelines to identify the time sequence of historical data.

3-4.3 (S1): Describe major factors that impact human populations and the environment.

Math

3.MD.B.3: Draw a scaled picture graph and a scaled bar graph to represent a data set with several categories. Solve one- and two-step "how many more" and "how many less" problems using information presented in scaled bar graphs.

3.MD.B.4: Generate measurement data by measuring lengths using rulers marked with halves and fourths of an inch. Show the data by making a line plot, where the horizontal scale is marked off in appropriate units—whole numbers, halves, or quarters.

(continued)

W—What Needs to Be Learned and Why?

WHY would the students care about this knowledge and these skills?

Craft the Driving Question that will lead to the development of the integrated tasks that provide for the application of the content, knowledge, and skills.

List the Essential Questions that can be answered as a result of the learning.

DRIVING QUESTION

How can we use concepts of force and motion to design a rover that can travel on all Martian surfaces?

ESSENTIAL QUESTIONS

Science

- How do forces make things move?
- How can we predict the motion of something?

Reading

- How do I use information in a text to explain the answer to a question?
- How can I use multiple ideas from a text to better understand what it's about?
- How does time, sequence, and cause/effect help us understand a text?

Writing

- How can I use writing to help me organize ideas?
- How can I write to communicate ideas clearly to others?

Social Studies

- How are humans impacting our environment and what are we going to do about it?

ESSENTIAL QUESTIONS

Math

- How can graphs make data more understandable?
- How can patterns in data explain phenomena?

H—How Do I Plan to Get There?

HOW do I plan to meet this goal?

Identify the pathway, including major tasks and milestones that result in answering the Driving Question.

1. Stomp Rocket activity
2. What Is Gravity? activity
3. Car Push Exploration activity
4. Source of Force activity
5. Wheel Size activity
6. Skype session with NASA's JPL engineers to discuss the Curiosity rover
7. Project: Rover will be "deployed" on the school playground while students remain in their classroom and view a live GoPro feed from a camera mounted on the rover. Students must program the rover to find, traverse, and investigate a possible "alien" life sighting.

E—What Evidence of Learning Will Be Used and How Will I Evaluate the Final Product or Project?

EVIDENCE and EVALUATE: What evidence of learning will be used and how will I evaluate the final product?

PRE-ASSESSMENT: What prior knowledge is needed for this task?

Identify the prerequisite skills and understandings.

- Complete district benchmark assessments for math, reading, and science.
- Review prior student work for writing.
- Check for vocabulary understanding, graphing skills, engineering process.

FORMATIVE: How will I measure student progress toward understanding?

Establish the assessment tools you will use to monitor progress and inform instruction.

Will use a variety of additional assessments (short answer, multiple-choice, open response) as needed to monitor student understanding of key science content, reading comprehension, application of key concept understandings.

SUMMATIVE: What criteria are needed for students to demonstrate understanding of the standards, content, and skills?

Create a checklist of criteria for use in a rubric.

Rubric sheet will be used as a tool to evaluate the individual group rover designs, including successful design parameters, analysis of the data for wheels and force, and overall success of mission in traversing the outdoor "Martian" landscape to find "alien" life.

R—How Will I Provide Opportunities to Increase the Rigor and Relevance?

RIGOR: How can I increase the student's cognitive thinking?

Identify tasks that can elevate student thinking, improve inquiry, and increase conceptual understanding.
Students will access knowledge from text and data collected in hands-on investigations. They will organize it and use it to inform their rover design plans; then they will build, test, and redesign their rovers.

RELEVANCE: Does the learning experience provide for relevant and real-world experiences?

Identify current topics and local issues that can make the tasks more engaging.

- How would living on Mars be different from living on Earth?
- Can I identify the forces acting on any given object?

E—How Do I Excite the Learners, Cognitively Engage Them, and Allow for Them to Explore for Deeper Understanding of the Content and Skills?

EXCITE: What is the hook to excite the learner?

Create the scenario to engage the learner.

Stomp rockets targeting a hula-hoop. This activity introduces the concepts of force and motion. It establishes the importance of good data and the concept of a fair test to obtain that data. It is hands-on, challenging, fun and can be referred back to throughout the unit.

ENGAGE: How will the students be cognitively engaged throughout the unit?

List the STEM practices that will be used as evidence.

How will the students be cognitively engaged throughout the unit?

Students will conduct investigations and develop models to gather, analyze, and interpret force and motion data.

They will use these data to develop a rover that is capable of traversing three terrain types found on Mars.

EXPLORE: What activities will help students address the Driving Question?

List questions for students to investigate that will lead them to a deeper understanding of the content and skills.

Stomp Rocket Activity

- Read and respond to "Forces Make Things Move" (comprehension and vocabulary questions).

Chair Push Activity

- Write to respond; balanced/unbalanced forces prompts.
- Explore tug-of-war.
- Engage in the four tug-of-war scenarios and diagram balanced and unbalanced forces for each.
- Read and respond to "Simple Physics of Soccer" (comprehension and vocabulary questions).
- "Neil Armstrong" text (ten chapters).
- Read one chapter a day and respond (power writing prompts; comprehension/vocabulary).
- Read and discuss "Gravity Is a Mystery."
- Read "Laws of Motion" (connect time, sequence, cause/effect in scientific process).

What Is Gravity? Activity

- Diagram force and motion on all three scenarios.
- Record data and create bar graph.
- Read "Wump World," view "Welcome to Mars" and "Asteroid Impact" videos.
- Discussion, "Why go to Mars?"
- Create posters representing reasons to explore Mars.
- Discuss "Habitat" and what it would take to live on Mars.
- Read article "Curiosity Rover Discovers a Perfect Home for Possible Life on Mars."
- Respond to comprehension questions.

Car Push Exploration Activity

- Record data on distance toy car traveled on various surfaces.
- Use data to predict future motion and sequence surfaces by distance traveled.
- Write to respond, "forces affecting (car's) motion."

Source of Force Activity

- Test and record data on the three sources (rubber band, ramp, balloon) of force for our rover.
- Graph data.
- Citing data, identify and predict most consistent source of force for rover.

(continued)

E

Wheel Size Activity
- Test and record data on distance traveled for three different wheel designs in all three Martian terrain types (rocky, flat, sandy).
- Graph data.
- Citing data, identify and predict best wheel design for each terrain type and best overall wheel design for our rover.
- Skype session with NASA's JPL engineers.
- Explore Curiosity rover.
- Generate questions for engineers.
- Skype with engineers and virtual tour of terrain test environment for Curiosity's earthbound twin.
- Students will build a solar-powered, remotely controlled rover from Lego Robotics club equipment.
- How do the forces of gravity and friction affect the motion of vehicles on different terrain types?
- How does wheel design affect a vehicle's motion on different terrain types?
- Why are we exploring Mars?
- What would it take to develop a human habitat on Mars?
- How does graphing our data make it easier to make decisions about our rover design?

Developing and Implementing the Unit

At Killip Elementary, we had implemented a building schedule that allowed an entire grade level of students to attend their special area classes at the same time. This allowed us to bring a grade-level group of teachers together at the same time for one hour every week. This time was used to strengthen curriculum, instruction, and assessment.

When it came time to develop a unit, we used these weekly collaborative team times to take a look at which standards would need to be addressed in the unit. With the standards then identified and some ideas on the table, each teacher spent a few days researching and identifying possible student resources that could allow us to teach the targeted standards. In the case of this Rover unit, much of the work had been done the previous year. We had already developed a third-grade unit with the appropriate texts on force and motion, gravity, friction, and some of the digital media necessary. However, we felt that this unit was heavy on science investigation and offered the students a very "hands-on" but fragmented learning experience.

We knew we wanted to align all of our instructional activities to one overarching Guiding Question but we didn't know what that should be or how to craft the intent of the unit. One member of the team had the idea for the Stomp Rocket activity as the "hook," but we struggled with the idea of how to make rockets personally or locally relevant to students. Through conversation around increasing the relevancy of the unit, the idea arose of doing something around the Curiosity rover that was currently operating on Mars. The idea was a hit and we set a date to bring in substitutes for the third-grade classes and spend the day with the team aligning the instructional activities for the unit to the Rover idea. Three months later, our students were sitting in front of the projector discussing their wheel designs of their own Curiosity rover with NASA's JPL engineers Laura Fisher and Hallie Gengl. It was exactly this kind of experience that we wanted to use as the foundation for our twenty-first-century curriculum. The experience engaged and motivated the students in their own learning while at the same time providing the school with a way to remain answerable to the standards-based accountability measures.

Although we are incredibly proud of this unit, we are currently looking at several possible revisions. We felt that the rover design and build activities became so engineering-heavy that we lost track of the science content. We also saw that need for explicit instruction and guidance with the twenty-first-century skills, particularly with communication and collaboration. We will include all of this next year, starting by developing a more detailed end-of-project rubric that includes specific indicators. We are grateful for an amazing staff of teachers, instructional coaches, and administrators that have allowed, supported, and driven our progress in STEM education. Our students are now being exposed to knowledge, skills, and opportunities they wouldn't have otherwise had access to. (See Figures 8.2, 8.3, and 8.4.)

FIGURE 8.2 Team C with Their Rover

FIGURE 8.3 Wheel Type Activity

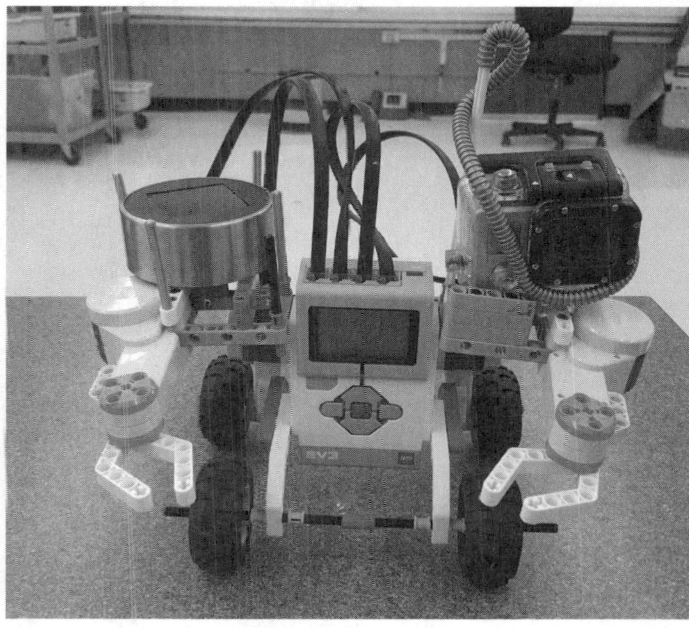

FIGURE 8.4 Final Rover Design

Think About It

Killip Elementary School's STEM journey began with the feeling of social responsibility to the members of their community, providing the kinds of foundational experiences that support long-term learning success. Through the sustained, long-term commitment of staff and administration, they have succeeded in changing the culture within their school community. By focusing on the explicit instruction of key skills and attitudes as needed, they were able to transform their school into a model of positive growth. Today, they continue in their desire to improve the opportunities for all of their students.

Speed Bumps

- How could you use the CBAM model to support your peers through the process of change?
- In what ways does your school or district resemble Killip Elementary? How are you different?
- What changes would you make to Killip Elementary's implementation model for your school or district?

Creating a Pathway

Fostering Community Collaboration

It is also possible to think about effective STEM schools in terms of different school types or programs that focus on STEM . . . Such schools and programs are important because they can serve as exemplars for districts across the nation that are attempting to elevate the quality of STEM education.

—National Research Council, *Successful K–12 STEM Education: Identifying Effective Approaches in Science, Technology, Engineering, and Mathematics*

In this chapter, we will look at Hillsborough, Florida, a large county public school district, whose approach to creating a systemic STEM initiative highlights the need to have all stakeholders involved in the process of developing and implementing a successful STEM program.

- First, they define what STEM is and define a shared vision and goal for their students, their teachers, and their communities.
- Second, they recognize the importance of the role of the teacher in making the initiative successful.

- Third, they worked to change the culture and mind-set of all the stakeholders, including parents, administrators, and members of the community by setting realistic expectations and determining shared responsibilities.

Dr. Larry Plank, Director of K–12 STEM Education, shares how their pathway to STEM teaching and learning evolved so that you may benefit from their experience as you develop your own STEM journey.

Hillsborough County Public Schools, Florida—Dr. Larry Plank

> *As I reflect upon our STEM journey over the past six years—the relationships, vantage points, and various lenses have all contributed to creating multiple levels of success in our district.* (Dr. Larry Plank, K–12 Director for STEM Education, Hillsborough County Public Schools, personal communication July 2016)

Our District Profile

Hillsborough County Public Schools (HCPS) is the eighth largest district in the nation, serving nearly 212,000 students at 250 school sites. The district is a microcosm of the nation, serving the city of Tampa, with inner-city issues of poverty and crime, suburban areas representing wealth, and rural areas comprised of both permanent families and thousands of migrant workers. The Port of Tampa is the area's largest economic driver. Tourism and the service industry fill the next largest portions of the economic engine, with advanced manufacturing, technology sector, and health care research painting the remainder of the financial picture and planned future of the region.

Beginning with What We Have

When we first began our journey six years ago, as the newly appointed K–12 Director for STEM Education, I was tasked with connecting a wide variety of departments in our district, from elementary education to career and technical education, magnet programs to out-of-school time programs, all while honoring the work that had been done previously and developing a collaborative vision and mission for STEM education for the foreseeable future. The work included over twenty district leaders, each with their own unique biases. We began the process by creating an asset map of what was working in our district, followed by what we wished we could become over time. The asset map helped to highlight the rich array of already existing materials and programs within the county that had been developed over time by various district leaders. This practice lessened the tensions among those who felt great ownership (and responsibility) for aspects of the work that had been done to date, and allowed for multiple leaders to come to grips with the fact that although well intentioned, our district portfolio was truly a collection of "random acts of STEM."

From this, we created a vision for STEM education in our district, along with four essential pillars for STEM education, and agreed that each future action of our district, no matter how big or small the endeavor, would fall into and support one or more of these pillars.

Essential Pillars for STEM Education

Essential Pillar 1: Curricular Innovations That Support Twenty-First-Century Skills and STEM Learning for All

Curricular innovations consist of changes to the standard curriculum, associated instructional practices, and district protocol that promote STEM programs and understanding and support learning in STEM subjects. At the center of this effort lies the implementation of the Florida Standards (based upon the Common Core State Standards) and the Next Generation Science Standards—both of which include practice standards that support the integrative STEM approach to curriculum. Additional instructional practices and tools to embrace our "STEM for ALL" vision, include but are not limited to: project-, problem-, and place-based learning approaches; engineering and design challenges; inquiry-based methodologies in mathematics and science; increasing enrollment and success of traditionally underrepresented groups in rigorous STEM coursework (such as Advanced Placement [AP]); and providing supports for students who engage in rigorous coursework.

The foundation for STEM success is built in the elementary grade-level bands, where engineering and design challenges have been incorporated into the formal curriculum, and mathematics and science "replacement" lessons from the core curriculum are made available to schools willing to take on newly designed methodologies immersed in the inclusion of robotics. In the middle grades, our district has focused upon integrating mathematics and science course content, where appropriate, and we have worked with a wide variety of agencies on multiple projects to create STEM lessons for grades 6–8. In high school, our efforts have centered upon preparing students for rigorous STEM coursework, with the goal of each student taking at least one AP course prior to graduation. Four examples of our work are described below.

1. *Elementary School Engineering and Design Challenges: Early, Successful First Steps.* During the 2012–13 school year, the elementary school STEM team unveiled a series of engineering and design challenges, which were incorporated into the formal curriculum of all grade levels (K–5) at all school sites for one hour per week. The lessons are focused upon real-life, relevant problems in the Tampa Bay community that afford students the opportunity to think critically and creatively to find solutions. The design challenges are interdisciplinary in nature to include mathematics, science, social studies, language arts, and visual arts. They are iterative in nature, requiring two to four weeks of time to complete. The content of each challenge is structured such that the subsequent grade level builds upon the previous year's big ideas, each with versatile constraints based upon grade level standards and student abilities.
2. *Elementary STEM Robotics Replacement Lessons: Technology for Impoverished Schools.* In 2011, nine schools were invited to participate in a project to develop anchor lessons in mathematics and science for elementary grades. These schools were designated as Title I and had challenges characteristic to similarly identified schools, such as teacher fatigue leading to high teacher turnover, discipline issues leading to a lack of student engagement, and a minimal amount of appropriate, usable technology for STEM teaching and learning. Each school principal was required to commit to supporting the use of robotics as a replacement for more traditional mathematics and science lessons, and the teachers were required to attend professional development, which consisted of book study sessions based upon innovative approaches in STEM, the use of robotics and associated programming, working with underserved populations, and finally writing and critiquing lessons. For their commitment, each school received a complete set of robots for use in fourth and fifth grades.

   Five years later the number of schools participating in this program has expanded to over forty, and most have enjoyed a more stable teaching force, decreased disciplinary issues, and an increase in science and/or mathematics scores on high stakes state assessments.

3. *Middle School STEM Integration Program: Learning to Collaborate and Support Integration.* Also during the 2011–12 school year, middle school mathematics and science teams at seventeen middle schools worked collaboratively to develop problem-based integrative STEM lessons. The project, entitled "STEM: Strong & Steady," was funded by a local education foundation and codirected by a leading state educational think tank and HCPS. We developed twelve problem-based lessons for mathematics and science classrooms (four per grade level), including interdependent components (i.e., data collected in science classrooms were analyzed in mathematics classrooms, or design principles were used in either room to solve discipline-specific problems). This project represented the first time our district required collaboration among discipline-specific teachers. We anticipated a dissonance between subject-area practices and frameworks for learning and spent the first year of the grant project defining STEM integration, building collaboration between mathematics and science departments in each school site, and finally preparing site administrators to support such collaboration in their schools. At the conclusion of the project four years later, the district had much more than twelve new lessons: we had created a model for integration that embraced inquiry-based instructional practices not only in mathematics and science, but also language arts; developed a platform for social and emotional education to include twenty-first-century learning skills such as collaboration and critical thinking; arrived at engineering as a touch point for all content areas; and began using design principles in our own work.
4. *AP STEM Access: Increasing Access and Equity to Advanced Coursework.* In 2012, the district sought to increase the number of traditionally underrepresented minority and female high school students that participate in AP courses in STEM disciplines through the AP STEM Access program. The project provided each participating school with start-up funding for the classroom resources, educational materials, and teacher professional development typically needed to start one or more new AP mathematics and science courses. As further incentive, all AP STEM teachers in participating schools that increased diversity in their classroom over the previous year's school rate received a $100 gift card for each student who achieved

a score of 3, 4, or 5 on an AP STEM exam. The gift cards are used by the teacher to further invest in classroom resources—with the goal of driving student engagement and achievement in years to come.

Essential Pillar 2: Establishing Career Pathways That Support Employment in STEM Fields

Although many schools and programs are focused upon postsecondary academic pathways for students in STEM, many overlook another true need in our country—STEM-literate, readily employable citizens who are prepared for the technological workforce. Although the nation currently enjoys a steadily decreasing unemployment rate, thousands of positions that require skills related to STEM are unfilled, costing the United States economy billions of dollars. The same can be said for greater Tampa Bay. In response, HCPS and our partners continue to fine-tune traditional programs in career and technical education (CTE) to address the needs of the nation and serve the community of students to whom we are responsible.

Our practices include:

- establishing programs of study that foster growth and understanding of relevant STEM content
- incorporating instructional models such as inquiry-based experiences and engineering/design challenges
- strengthening connections to local industry
- adopting best practices through Career Academy models
- employing appropriate support structures for students in CTE programs.

When we first set out on our collective STEM mission, STEM-themed programs in CTE were scattered throughout our district, without a coherent pathway between them. For instance, a middle grades engineering program might reside twenty or more miles from the high school equivalent, and the local high school might have an agriscience program. In addition, many programs were out of date and in need of a curricular upgrade.

The district prioritized the development of new and connected STEM programs in middle and high school, and in 2011 ten new STEM schools emerged that utilized a career academy model in that all students were admitted as cohorts and continued through the program together. Middle and high school articulation was established, and as a result, the district celebrated the birth of connected pathways in STEM: robotics, engineering, biotechnology, and aerospace. In addition, the district added six new magnet schools with STEM themes.

Teachers within the programs report to school one week earlier than their peers and are also provided with common planning time during the school day. The student experience includes connected, integrated learning enhanced by multiple field learning days at related business and industry partners.

Essential Pillar 3: Fostering Community Relationships That Support STEM Learning

Many school districts are concerned with making connections to the home to ensure parental support for students in the educational process. However, establishing parental connections with STEM-related programs can be more difficult than in other areas, because parents may not have great familiarity with the STEM fields. To aid in the effort to bridge communication and understanding gaps, HCPS has partnered with the Alliance for Public Schools, which host Parent University and other family-centric events throughout the year. In addition to parents, we actively seek support from academic, business, and community partners, which we refer to as the "ABCs of STEM" and are essential to STEM programs because public schools rely upon these institutions for innovations within the STEM fields, financial backing, and academic guidance and support. HCPS has a long-standing relationship with two major universities (University of South Florida and University of Tampa) and a large community college system (Hillsborough Community College).

Essential Pillar 4: Value-Added and Nontraditional Programs That Support STEM Learning

Research suggests that much of what students learn in STEM disciplines, especially science, occurs through discovery and exposure to content outside of the typical classroom. This learning can occur through self-guided exploration, experiences at informal science institutions, as well as through a variety of media. We use programs such as after-school extended learning programs, Saturday school, competitions, fairs, and other community events to bring content to life for students. HCPS fostered the growth of our STEM programs through long-standing partnerships for teachers and students with the Museum of Science & Industry and the Florida Aquarium; after-school programs that support STEM-related clubs, ranging from environmental stewardship to robotics; special programs, including "Saturdays with STEM" and "STEM Goes to Work"; and the largest regional STEM Fair in the state supported by the Hillsborough Education Foundation.

A Sustainable Future: The STEM Ecosystem

Educators by nature feel a deep-rooted responsibility to ensure that all students are successful. This passion to support future generations by creating engaging learning spaces in the schoolhouse and connecting with parents to keep students on a trajectory toward a successful life resonates with educators across the nation regardless of community. Although technology has played a role in connecting classrooms to the outside world, teaching and learning has remained a very personal, closed experience in many schools and districts.

What we have learned in Hillsborough County is that purposeful collaboration between multiple sectors in the community—businesses, STEM-rich informal learning institutions, parents, and formal educators—results in an increased level of success for students and the promise of well educated, STEM-literate, community members.

Action Steps—From the Authors

Now that we have shared the STEM lesson guideposts as a tool for planning STEM units, we've summarized the key elements to successful implementation of STEM approaches within a system.

- *First and foremost, a supportive administration that will help provide the time for teacher collaboration and professional development*
    a. Set up a time to discuss this planning process and model with your administration.
    b. Identify collaboration needs and review scheduling with administration.
    c. Set a professional development schedule for all teachers to further understand the STEM lesson guideposts and the process of unit planning.
- *Access to the resources needed for curriculum and supplies*
    a. Use the W.H.E.R.E. Model and STEM Lesson Guidepost Planning Template to develop your own STEM unit using your existing curriculum.
    b. Identify existing resources and those that you may need, including technology, materials, and people.
- *A good system of communication including parents, caregivers, and community members describing what STEM learning is and how its benefits will affect their children as they apply their skills and knowledge in new and different ways.*
    a. Plan a STEM family night with rigorous and relevant STEM activities that families can participate in together.

b. Share the benefits of STEM teaching and learning with parents at conferences and in district communications (newsletters, school website, and so on).

c. Use parent volunteers to lend expertise on STEM subjects and career connections throughout the year and across the district.

- *Establishing community partnerships to help support teachers and provide supporting resources and expertise*

  a. Find community partners to lend expertise on STEM topics, issues, and career connections.

  b. Contact existing community organizations such as the chamber of commerce, rotary clubs, and business/education committees in your local area for support.

Good luck as you embark on your own STEM journey!

The journey of a thousand miles begins with one step. (Lao Tzu)

Think About It

Every district has its own unique set of opportunities and challenges and Hillsborough County, Florida, is no exception. Bringing about change began with a process for developing an asset map, identifying what was already created and available, followed by what district members wished they could accomplish over time, and resulted in the creation of their needs list. The resulting multitiered approach to implementing a comprehensive and cohesive STEM policy combined many facets of the community, bringing together internal and external resources and members of the community under a shared vision for STEM integration.

Speed Bumps

- In what ways do your district challenges resemble those of Hillsborough County? How are you different?
- How might your district move to a more inclusive and coherent STEM policy?
- What local and regional resources could you take advantage of to further the implementation of a cohesive STEM program in your district?

Appendix A

Whirligig Pattern

(Cut on solid lines, fold on dotted lines.)

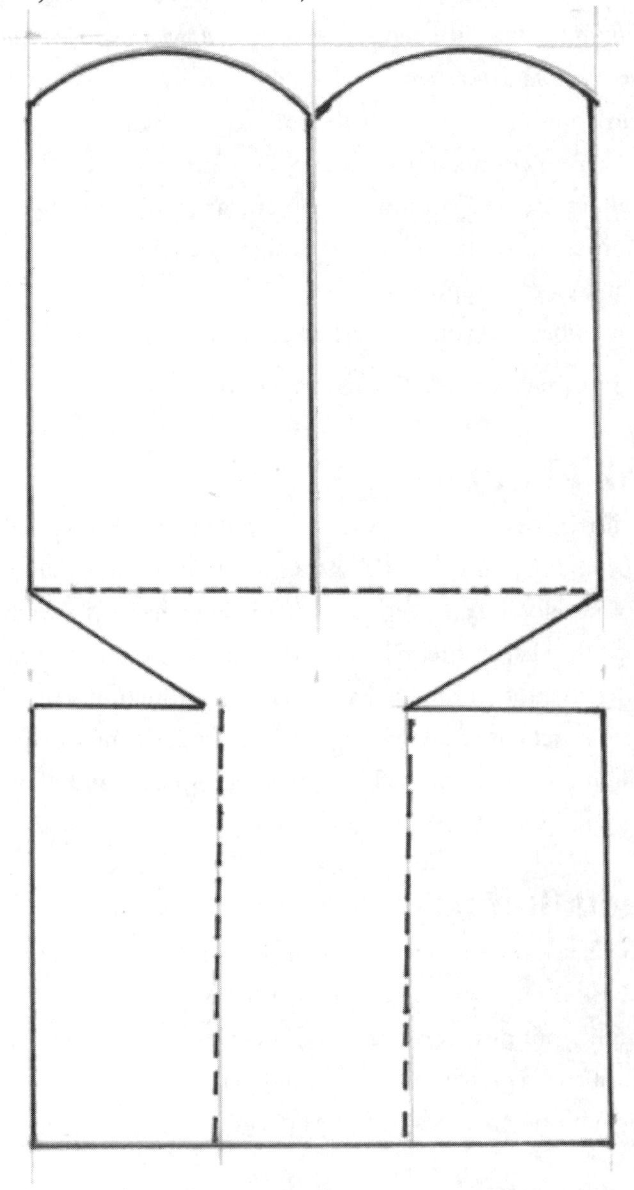

Appendix B

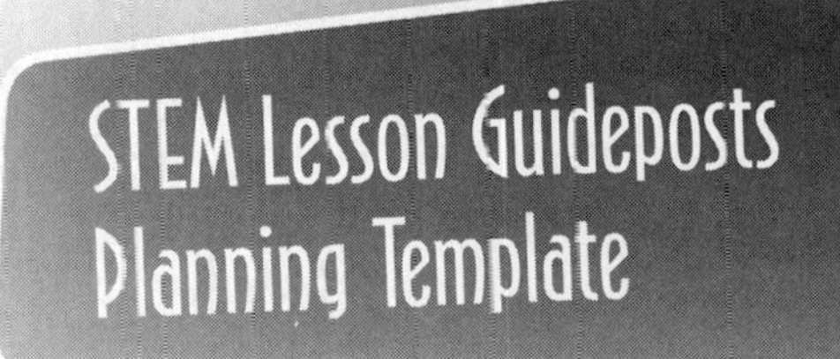

WHAT are the desired results, including big ideas, content standards, knowledge, and skills?

List the content standards and what the students will know and be able to do.

WHY would the students care about this knowledge and these skills?

Craft the Driving Question that will lead to the development of the integrated tasks that provide for the application of the content, knowledge, and skills.

List the Essential Questions that can be answered as a result of the learning.

HOW do I plan to meet this goal?

Identify the pathway, including major tasks and milestones that result in answering the Driving Question.

EVIDENCE and **EVALUATE:** What evidence of learning will be used and how will I evaluate the final product?

PRE-ASSESSMENT: What prior knowledge is needed for this task?

Identify the prerequisite skills and understandings.

FORMATIVE: How will I measure student progress toward understanding?

Establish the assessment tools you will use to monitor progress and inform instruction.

SUMMATIVE: What criteria are needed for students to demonstrate understanding of the standards, content, and skills?

Create a checklist of criteria for use in a rubric.

RIGOR: How can I increase students' cognitive thinking?

Identify tasks that can elevate student thinking, improve inquiry, and increase conceptual understanding.

RELEVANCE: Does the learning experience provide for relevant and real-world experiences?

Identify current topics and local issues that can make the tasks more engaging.

EXCITE: What is the hook to excite the learner?

Create the scenario to engage the learner.

ENGAGE: How will the students be cognitively engaged throughout the unit?

List the STEM practices that will be used as evidence.

EXPLORE: What activities will help students address the Driving Question?

List questions for students to investigate that will lead them to a deeper understanding of the content and skills.

Appendix C

STEM Lesson Guideposts Planning Template

Engineering a Hat

W

WHAT are the desired results, including big ideas, content standards, knowledge, and skills?

List the content standards and what the students will know and be able to do.

Students will design a hat that can be worn in the hot, sunny conditions but also have a way to keep them cool. They will compare similar designs, describe the shapes used, and discuss the advantages/disadvantages of the design for the type of weather conditions it will be used for.

MAIN STANDARD (Driver)
Engineering Design
ETS-1: Defining and delimiting engineering problems
ETS-2: Developing possible solutions
ETS-3: Optimizing the design solutions

WHY would the students care about this knowledge and these skills?

Craft the Driving Question that will lead to the development of the integrated tasks that provide for the application of the content, knowledge, and skills.

List the Essential Questions that can be answered as a result of the learning.

DRIVING QUESTION
How can we, as engineers, design a hat that will protect us from the hot sun and also keep us cool?

ESSENTIAL QUESTIONS
Engineering Design
- What observations and questions need to be answered before you can solve the problem?
- How can the engineering design process be used to find a solution to the problem?

H

HOW do I plan to meet this goal?

Identify the pathway, including major tasks and milestones that result in answering the Driving Question.

Time to completion: 8 class periods
Class 1: Introduce the scenario, Driving Question, engineering design process.
Class 2: Describe how they will use this process to design their hats. Elicit the "needs to know."
Class 3–4: Research location, landscape, and weather of Grand Canyon.
Class 5: Develop the blueprint for approval of their hat design.
Class 6: Construct their hats and develop their explanation for how it solves their problem.

E

EVIDENCE and **EVALUATE:** What evidence of learning will be used and how will I evaluate the final product?

PRE-ASSESSMENT: What prior knowledge is needed for this task?

Identify the prerequisite skills and understandings.

Demonstrate understanding of two and four equal shares, different sizes, shape names, and attributes.

Define different terrains, mountains, rivers, valleys, and deserts.

Describe different types of weather conditions they have experienced. How are they different? How are they alike?

R

RIGOR: How can I increase the student's cognitive thinking?

Identify tasks that can elevate student thinking, improve inquiry, and increase conceptual understanding.

- Investigate the weather conditions in other places of their choosing—decide on the best design for their hat, which they can wear in both climates.
- Develop a blueprint plan for the construction of their hat, which will need to fit both types of weather conditions. And plan the sequence of steps for the hat's construction.
- Design a group presentation for explaining the construction and versatility of their hat.

E

EXCITE: What is the hook to excite the learner?

Create the scenario to engage the learner.

The *Engineering a Hat* STEM Unit Scenario: "The students will imagine that they are going on a summer driving trip with their family from Phoenix to the Grand Canyon. While visiting the Grand Canyon they will have a few days to walk around the rim of the Canyon and perhaps even hike partway down one of the trails. They are very excited and need to gather some information for their visit. First of all they will need to know what the weather is going to be during their stay."

ENGAGE: How will the students be cognitively engaged throughout the unit?

List the STEM practices that will be used as evidence.

SECONDARY STANDARDS (Passengers)

Science

ESS2.D: K–2 Weather and Climate

Weather is the combination of sunlight, wind, snow or rain, and temperature in a particular region at a particular time. People measure these conditions to describe and record weather and to notice patterns over time.

Social Studies

Strand 4: Geography

PO3: Discuss physical features (e.g., mountains, rivers, deserts) in the world. (With a connection to science: measure and record weather conditions, identify clouds and analyze their relationship to temperature and weather patterns.)

Mathematics

Geometry, reason with shapes and their attributes

2.G.A.3: Partition circles and rectangles into two, three, or four equal shares using the words *halves, thirds, half of, third of,* and so on and describe the whole as two halves, three thirds, four fourths. Recognize that equal shares of identical wholes need not have the same shape.

- How will drawing a picture or creating a blueprint of the design help with the construction of the hat?
- Is there more than one design that will provide for the best solution to the problem?

Science

- How can we find out about the weather at the Grand Canyon?
- What is the best tool for measuring the temperature?
- Why is there a difference between the nighttime and daytime temperature at the Grand Canyon?

Mathematics

- What shapes can be used to make our hat?
- How can shapes be divided into equal parts to make our hat look even?
- What are the attributes of the different shapes that are used in my hat?

Social Studies

- What is the landscape like at the Grand Canyon?
- How do the land features affect the climate of the Grand Canyon?
- Why are there differences in temperature between the top and bottom of the Grand Canyon?

Class 7–8: Present hats to classmates. Discuss suggestions from classmates on how designs can be changed or better "optimized" to solve the problem.

FORMATIVE: How will I measure student progress toward understanding?

Establish the assessment tools you will use to monitor progress and inform instruction.

Checking individual understanding using a individual student checklist rubric.

- Engineering design process
- Identification of landform features
- Relationship between temperature and weather
- Partitioning of shapes

SUMMATIVE: What criteria are needed for students to demonstrate understanding of the standards, content, and skills?

Create a checklist of criteria for use in a rubric. Students are scored on: the hat's design, protection for sun; cooling effect; use of shapes, and application of the engineering design process.

Individual response worksheets showing the process they followed, how they optimized their design, and their explanation of the shapes they applied in the construction of their hat.

What article of clothing would you like to design for our trip?

RELEVANCE: Does the learning experience provide for relevant and real-world experiences?

Identify current topics and local issues that can make the tasks more engaging.

Hats are a common pieces of clothing for students.

What do hats look like in other places in the world?

Apply their knowledge of hats to a real-world situation.

Have a clothing designer/engineer talk about they come up with new designs?

Practices and skills the students will be using:

- Develop and use a model
- Construct explanations and design solutions
- Engage in argument from evidence
- Use appropriate tools strategically

EXPLORE: What activities will help students address the Driving Question?

List questions for students to investigate that will lead them to a deeper understanding of the content and skills.

- How does the type of weather affect the design of a practical hat?
- When developing the blueprint, what directions will be needed so that any member of the team can construct the hat? Are they clear? Is the drawing accurate enough for anyone to follow?

Materials: straws, paper plates, yarn, tape, measuring tapes, scissors, cotton balls, different pieces of fabric; string, ribbons

- Is any member of the team, if called upon, able to fully explain the design and practicality of the group's hat?

Appendix D

Wind Turbine

STEM Lesson Guideposts Planning Template

W

WHAT are the desired results, including big ideas, content standards, knowledge, and skills?

List the content standards and what the students will know and be able to do.

The students develop an understanding of the relationship between changes in kinetic energy and the energy output of a system. They discover that the electrical output can be manipulated by design changes in the system.

By developing a model wind turbine to transfer electrical output data, students produce evidence that changes in kinetic energy are transferred from one system to another.

MAIN STANDARD (Driver) Science

MS-PS3-5: Construct, use, and present arguments to support the claim that when kinetic energy of an

WHY would the students care about this knowledge and these skills?

Craft the Driving Question that will lead to the development of the integrated tasks that provide for the application of the content, knowledge, and skills.

List the Essential Questions that can be answered as a result of the learning.

DRIVING QUESTION

As the Wind Turbine Power Company representative, present evidence that addresses the question, "Which blade design considerations transfer the greatest amount of kinetic energy?"

ESSENTIAL QUESTIONS

- How is kinetic energy transferred between objects or systems?
- How can kinetic energy in a system be measured?

H

HOW do I plan to meet this goal?

Identify the pathway, including major tasks and milestones that result in answering the Driving Question.

1. Small groups or students will work on constructing the turbine first, before more thorough investigation on given parameter. Will assess student proficiency after first class, and regroup student teams as necessary.
2. Review the initial guidelines for the turbine challenge, discuss the investigation constraints, and review the criteria on the evaluation rubric.
3. Three milestones: initial construction success; generation of the team's "variable" data; final presentation.

E

EVIDENCE and EVALUATE: What evidence of learning will be used and how will I evaluate the final product?

PRE-ASSESSMENT: What prior knowledge is needed for this task?

Identify the prerequisite skills and understandings.

Probe for student understanding on:

- energy transfer
- engineering/design process

Need to know:

- kinetic energy
- energy transfer
- motion
- force
- systems and subsystems
- evidence
- variables

FORMATIVE: How will I measure student progress toward understanding?

R

RIGOR: How can I increase students' cognitive thinking?

Identify tasks that can elevate student thinking, improve inquiry, and increase conceptual understanding.

Ask students to generate a list of blade parameters to investigate.

What other parameters in the wind turbine design could we explore to address claims that changes in kinetic energy result in changes of energy in or out of the system?

The ultimate solution is combining the individual parameters into a single design.

Allow students to determine which combination of different design parameters could produce the most energy transfer.

E

EXCITE: What is the hook to excite the learner?

Create the scenario to engage the learner.

The town is soliciting bids for an efficient wind energy generator built on a research-based design. The generator will be placed on the hill in the center of town, which gets a constant flow of westerly wind. Present a proposal for an efficient wind turbine design based on resarch. Contract award will be based on design research and committee presentation.

ENGAGE: How will the students be cognitively engaged throughout the unit?

List the STEM practices that will be used as evidence.

- Engage in argument from evidence.
- Analyze and interpret data.
- Plan and carry out investigations.

object changes, energy is transferred to or from that object.
DCI #1: When the motion of an object changes, there is inevitably some other change in energy at the same time.
DCI #2: When two objects interact, each one exerts a force on the other that can cause energy to be transferred to or from the object.

SECONDARY STANDARDS (Passengers)
Engineering/Design
ETS1-4: Develop a model to generate data for iterative testing and modification of a proposed object, tool, or process such that an optimal design can be achieved.

Mathematics
6.EE: Represent and analyze quantitative relationships between dependent and independent variables.
6.G: Solve real-world problems involving area, surface area, and volume.

ELA
W.6.1.A: Introduce claim(s) and organize the reasons and evidence clearly.
W.6.4: Produce clear and coherent writing in which the development, organization, and style are appropriate to task, purpose, and audience.

- How can the kinetic energy in a system be improved?
- Does the blade configuration, length, material, or angle impact the amount of kinetic energy transferred?
- How can the surface area of the blades be claculated?
- How do changes in the model wind turbine design affect its electrical output?
- How can I use the data I generate to support my claim?

4. Proposed five to seven class periods to accomplish:
Class 1: Build turbine and make modifications.
Class 2–4: Small-group investigations of given parameters; finalize report.
Class 5–7: Present oral and written reports.

Establish the assessment tools you will use to monitor progress and inform instruction.

Individual: Open response questions (five) about the engineering/design process and the evidence produced from the initial wind turbine building activity.
Intervention: Monitor student success after initial build; reassign groups if necessary to accommodate for special needs (motor skills/social skills).

SUMMATIVE: What criteria are needed for students to demonstrate understanding of the standards, content, and skills?

Create a checklist of criteria for use in a rubric.

Group: Will create an eight- to twelve-slide PowerPoint detailing procedure, results, conclusion.

Individual: Will write a one-page paper responding to the claim following the claim-evidence-reason format.

Rubric: Two rubrics needed; (refer to ELA standards for presenta tion/writing skils)

RELEVANCE: Does the learning experience provide for relevant and real-world experiences?

Identify current topics and local issues that can make the tasks more engaging.

To have students see the economic opportunity or career potential, have a wind turbine installer visit the class to describe the work involved in selecting and installing wind turbines.

To understand how your town government operates, have students research the actual bid process.

How does the town bid process work? What factors are needed to make up a competivie bid?

EXPLORE: What activities will help students' address the Driving Question?

List questions for students to investigate that will lead them to a deeper understanding of the content and skills.
- Do more blades make for more electrical output?
- Are longer blades more effective for transferring the kinetic energy of the wind?
- Is there a difference in energy output depending on the way the blade is angled on the spoke?

Materials: oak tag, plastic bottles, window fan

Reading: chapter text on energy transfer

Appendix E

Feeding Fluffy

STEM Lesson Guideposts Planning Template

W

WHAT are the desired results, including big ideas, content standards, knowledge, and skills?

List the content standards and what the students will know and be able to do. Strategies for measuring volume. Students will understand concepts of volume and relate volume to multiplication and to addition. Students will find volumes of right rectangular prisms. Students will understand that volume is additive.

MAIN STANDARD (Driver) Mathematics

5.MD.5b: Apply the formulas $V = (l)(w)(h)$ and $V = (b)(h)$ for rectangular prisms to find volumes of right rectangular prisms with whole-number edge lengths in the context of solving real-world and mathematical problems.

WHY would the students care about this knowledge and these skills?

Craft the Driving Question that will lead to the development of the integrated tasks that provide for the application of the content, knowledge, and skills.
List the Essential Questions that can be answered as a result of the learning.

DRIVING QUESTION

How can we use concepts of volume and simple machines to engineer a feeder for a given dangerous animal?

ESSENTIAL QUESTIONS

- How is volume related to the size and shape of an object?
- What is the relationship between area and volume?
- How does the area of a rectangle help us find the volume of a rectangular prism?

H

HOW do I plan to meet this goal?

Identify the pathway, including major tasks and milestones that result in answering the Driving Question.
Class 1: Launch task—feeding neighbor's Fluffy
Class 2: Exotic animals as pets argumentation
Class 3: Animal research—type and amount of foods different animals eat
Class 4: Exploring volume—packing unit cubes; use volume formula
Class 5: Container designs—use concepts of volume to design a container
Class 6: Forces and simple machines exploration

E

EVIDENCE and EVALUATE: What evidence of learning will be used and how will I evaluate the final product?

PRE-ASSESSMENT: What prior knowledge is needed for this task?

Identify the prerequisite skills and understandings.

Linear measurement using standard and nonstandard units

Finding the area of a rectangle using square units and multiplicative thinking

Determining the relationship between area and perimeter

FORMATIVE: How will I measure student progress toward understanding?

Establish the assessment tools you will use to monitor progress and inform instruction.

Packing unit cubes (nonstandard) to fill a rectangular prism (performance-based)

R

RIGOR: How can I increase students' cognitive thinking?

Identify tasks that can elevate student thinking, improve inquiry, and increase conceptual understanding.

The following tasks provide students an opportunity to extend their thinking. Change the scenario to include designing a food container for a grizzly bear.

- Students research how much food a grizzly bear eats.
- Calculate how much space that amount of food takes up.

Students are given the task of designing a food delivery system that can move the animal's food container using a variety of simple machines.

- Students must figure out which system of simple machines can work together to deliver the container.

Designing and presenting their final products forces students to apply the knowledge and skills from multiple disciplines and provide evidence to justify their designs.

E

EXCITE: What is the hook to excite the learner?

Create the scenario to engage the learner.

Students are presented with the scenario of a neighbor leaving town and they will need to feed the neighbor's pet, Fluffy, while the neighbors are gone, but the neighbors forgot to leave a container. Students will discuss constraints and other factors related to feeding Fluffy.

ENGAGE: How will the students be cognitively engaged throughout the unit?

List the STEM practices that will be used as evidence.

Reason abstractly and quantitatively

Create and use models

Ask questions and define problems

SECONDARY STANDARDS (Passengers)

Science

S5-C2-PO 3: Examine forces and motion through investigations using simple machines (e.g., wedge, plane, wheel and axle, pulley, lever).

ELA

5.R.I.9: Integrate information from several texts on the same topic to write or speak about the subject knowledgeable.

5.W.9: Draw evidence from literary or informational texts to support analysis, reflection, and research.

- How do simple machines help us do work?
- What is the relationship between force and motion?

Class 7: Compound machine weight movement task
Class 8: Sketching the solution
Class 9: Evidence; justification
Class 10: Presentation

Counting standard cubic units that fill a rectangular prism (performance-based or district assessment)

Label the dimensions of a rectangular prism filled with standard cubic units (written formative or district assessment)

Use volume formula to determine the volume of a rectangular prism given its dimensions, or find a missing dimension given the total volume and two dimensions (written formative or district assessment)

SUMMATIVE: What criteria are needed for students to demonstrate understanding of the standards, content, and skills?

Create a checklist of criteria for use in a rubric.
Students will be able to design a container that holds the amount of food for their given animal.

The design blueprint with correctly labeled dimensions can be used as a summative assessment. A blueprint design rubric will be used listing criteria to be included that provides evidence that students used concepts of volume and simple machines and applied them correctly to the development of their container and feeder.

RELEVANCE: Does the learning experience provide for relevant and real-world experiences?

Identify current topics and local issues that can make the tasks more engaging.

Students will take the role of engineers to design an animal feeder for dangerous animals.

- This brings relevance by connecting the learning to work done in the real world by engineers when they design solutions to problems.
- Discussions on raising animals in captivity and keeping dangerous animals as pets helps students vest themselves in the tasks and learning because these topics relate to real-world situations.

They will use the concepts of volume to design a container that holds the necessary amount of food.

- Any time you apply concepts and skills to solve a problem or fill a need, the relevance of the learning is at its hightest level.
- The type of animal and what it eats provides relevance to the volume and the size of the container needed.

They will use the concepts of force, motion, and simple machines to make a machine that can lift, transport, and deliver the food to their designated animal.

- Again, application of skills and concepts as they are being learned keeps the students connected to the learning and avoids it being disjointed from other concepts and skills.

EXPLORE: What activities will help students address the Driving Question?

List questions for students to investigate that will lead them to a deeper understanding of the content and skills.

Design a container to feed Fluffy: What is volume?

Creating prisms: How does the idea of area help to develop the idea of volume?

Joel's Buildings: How is volume related to the size and shape of an object?

Design a feeding system for a given dangerous animal: How is volume related to the size and shape of an object? How do simple machines help us do work? What is the relationship between force and motion?

Materials: centimeter cubes, grid paper, simple machine sets, poster paper, markers

Resources: zoo animal feeding video clips from YouTube

Exploration Tasks for *Feeding Fluffy*

DRIVING QUESTION

How do we use concepts of volume and simple machines to engineer a feeder for a given dangerous animal?

1. LAUNCH TASK—FEEDING NEIGHBOR'S FLUFFY

- **Design a bowl to feed neighbor's Fluffy**
  - **Essential Questions:** How big of a container is needed to feed a grizzly bear? Do different foods take up different amounts of space? Do ninety pounds of salmon require the same size container as ninety pounds of berries?
  - **Anticipated/targeted STEM practices:** When students are presented with the problem and are asked to brainstorm a design they will ask questions, define problems, make sense of problems, and persevere. When they share their design concepts with the whole group, students will also ask questions and construct viable arguments.
  - **Rigor level:** The students are at the strategic thinking level, which requires reasoning, developing plan or sequence steps, some complexity, and more than one possible answer.
  - **Materials:** markers, poster paper, research materials or Web access.

2. EXOTIC ANIMALS AS PETS ARGUMENTATION

- **Socratic seminar**
  - **Essential Questions:** Should dangerous animals be allowed to be kept as pets? What is the difference between captivity in zoos versus captivity as pets?
  - **Anticipated/targeted practices:** When students look up data for evidence to justify their opinion they will build strong content knowledge, respond to the varying demand of audience, and they will value evidence. When they engage in discussion through a Socratic seminar, they will comprehend as well as critique, and come to an understanding of other perspectives.
  - **Rigor level:** Students will be at the strategic thinking level, which requires reasoning and developing plan or sequence steps, and is evidenced by students assessing, citing evidence, critiquing, developing a logical argument, differentiating, and drawing conclusions.
  - **Materials:** note cards, research materials or Web access, pencils, note journals.

3. ANIMAL RESEARCH—TYPE AND AMOUNT OF FOODS DIFFERENT ANIMALS EAT

- **Animal research chart**
  - **Anticipated/targeted practices:** As students gather information about their animal to place on their chart, they will be building strong content knowledge and valuing evidence.

- **Rigor level:** Recall reproduction stage because at this time they are just gathering information and tabulating it. They are not being asked to reason with it in this task, but the information will be connected to a future task that requires them to reason with this information.
- **Materials:** poster paper, markers, rulers, research materials or Web access.

4. EXPLORING VOLUME—PACKING UNIT CUBES; USE VOLUME FORMULA

- **Creating prisms activity**
  - **Essential Questions:** What is volume? What is the relationship between the square unit and the length and width dimensions of the rectangle? How can we use standard units of measurement to find the volume of a rectangular prism? How do we stack standard cubic units to find the volume of a rectangular prism?
  - **Anticipated/targeted practices:** Students will be creating rectangles with square unit tiles, and then rectangular prisms by stacking unit cubes and counting. This will get the students to reason abstractly and quantitatively, and construct viable arguments as they share their created prisms.
  - **Rigor level:** Students will be at the skill/concept level as they are still building some knowledge. They are engaging in mental process beyond habitual response using information or conceptual knowledge.
  - **Materials:** centimeter grid paper, rulers, pencils, centimeter cubes.
- **Joel's Buildings Activity**
  - **Essential Questions:** Can we multiply the length by the width and by the height to find the volume of a rectangular prism?
  - **Anticipated/targeted practices:** This transition activity bridges students' conceptual understandings of volume to the volume formula so they will be modeling with mathematics, reasoning abstractly and quantitatively, looking for and making use of structure, using appropriate tools strategically.
  - **Rigor level:** Students will be at the strategic thinking level because this requires reasoning and developing a plan or sequence steps.
  - **Materials:** centimeter grid paper, rulers, pencils, centimeter cubes.

5. CONTAINER DESIGNS—USE CONCEPTS OF VOLUME TO DESIGN A CONTAINER

- **Container blueprints**
  - **Essential Questions:** How big of a container is needed to feed a given animal? How are the dimensions of rectangular prism labeled on a blueprint?
  - **Anticipated/targeted STEM practices:** When students are creating the design and blueprint of their container they will make sense of problems and persevere, and reason abstractly and quantitatively. They will model with mathematics by generalizing a formula that can be applied in similar situations. When they share their design blueprints with the whole group, students will also ask questions and construct viable arguments.

- **Rigor level:** The students are at the extended thinking level because designing the container given a set of parameters requires investigation, complex reasoning, planning, and developing.
- **Materials:** poster paper, markers, notes journals.

6. FORCES AND SIMPLE MACHINES EXPLORATION

- **Building simple machines with K'Nex connector sets**
  - **Essential Questions:** What are simple machines? How do simple machines help us do work?
  - **Anticipated/targeted STEM practices:** When students are making simple machines with the connector sets, they are asking questions, planning and carrying out investigations, and constructing explanations about simple machines and how they work.
  - **Rigor level:** Students will be at the strategic thinking level, which requires reasoning and developing a plan or sequence steps, and is evidenced by students assessing, citing evidence, critiquing, developing a logical argument, differentiating, and drawing conclusions.
  - **Materials:** various simple machines, K'Nex connector sets.

7. COMPOUND MACHINE WEIGHT MOVEMENT TASK

- **Weight movement task**
  - **Essential Questions:** What is a force? What is motion? What is the relationship between force and motion? How can simple machines work together to make a compound machine?
  - **Anticipated/targeted STEM practices:** When students design a compound machine they will define problems, plan and carry out investigations, analyze and interpret data, and design a solution to move a weight using a combination of simple machines.
  - **Rigor level:** The students are at the extended thinking level because designing a working compound machine requires investigation, complex reasoning, planning, and developing, while creating and applying concepts.
  - **Materials:** various simple machines, K'Nex connector sets, gram weights, cups, tape, string, paper clips.

The final three tasks are culminating products of learning and integrating multiple disciplines at the application level. The Essential Questions for learning have already been addressed through previous tasks, and these are the final tasks that address the Driving Question listed previously.

8. SKETCHING THE SOLUTION

- **Design blueprint for automatic feeding system**
  - **Anticipated/targeted STEM practices:** When students design their automatic feeder and sketch their blueprint they will model with mathematics, define problems, and design solutions.
  - **Rigor level:** The students are at the extended thinking level because designing and sketching an automatic feeding system requires investigation, complex reasoning, planning, and developing, while creating and applying concepts.
  - **Materials:** poster paper, markers, computers, PowerPoint or other presentation software.

9. EVIDENCE; JUSTIFICATION

- **Socratic seminar for design critique**
  - **Anticipated/targeted practices:** When students use evidence to justify their design to their peers, they will respond to the varying demand of audience, and they will value evidence. When they engage in discussion through a Socratic seminar, they will comprehend as well as critique, and come to an understanding of other perspectives.

- ◦ **Rigor level:** Students will be at the strategic thinking level, which requires reasoning and developing a plan or sequence steps, and is evidenced by students assessing, citing evidence, critiquing, developing a logical argument, differentiating, and drawing conclusions.
- ◦ **Materials:** projector, computer, posters, note cards, journals.

10. PRESENTATION

- **Zoo officials panel presentation.**
- **Anticipated/targeted practices:** When students use evidence to explain their design to an expert panel of zoo officials, they will respond to the varying demand of audience, and they will value evidence. When they engage in discussion through a Socratic seminar, they will comprehend as well as critique, and come to an understanding of other perspectives.
- **Rigor level:** Students will be at the recall/reproduction level as they recall facts and information about their design and the procedures of how it works. They will arrange their information and use it to tell what, when, and why and how about their design.
- **Materials:** note cards, journals, multimedia presentations.

Art Pedestal Project

STEM Lesson Guideposts Planning Template

W

WHAT are the desired results, including big ideas, content standards, knowledge, and skills?

List the content standards and what the students will know and be able to do.

Define a simple design for a problem that reflects the criteria for success and constraints on materials, time, and cost.

MAIN STANDARD (Driver)
D2.Civ.7.6-8: Apply civic virtues and democratic principles in school and community settings. (The virtues cited are "honesty, mutual respect, cooperation and attentiveness to multiple perspectives that citizens should use when they interact with each other on public matters.")

Twenty-first century skills:

- Effective communication
- Teamwork
- Collaboration
- Interactive communication

WHY would the students care about this knowledge and these skills?

Craft the Driving Question that will lead to the development of the integrated tasks that provide for the application of the content, knowledge, and skills.

List the Essential Questions that can be answered as a result of the learning.

DRIVING QUESTION
How can our team build a structure to support this famous statue, which is attractive, built within the budget, and finished on time?

ESSENTIAL QUESTIONS

- How does a community make decisions?
- What observations and questions need to be answered before creating the design for the pedestal?

H

HOW do I plan to meet this goal?

Identify the pathway, including major tasks and milestones that result in answering the Driving Question.

Class 1: Background building—Show several different commonly known statues beginning with the Statue of Liberty. Discuss statues. Discuss how these statues are supported. Introduce the scenario and pose the Driving Question. Introduce or review the engineering design process and discuss how this will help students to build their art pedestal. Record their "needs to know" questions on chart paper.

E

EVIDENCE and EVALUATE: What evidence of learning will be used and how will I evaluate the final product?

PRE-ASSESSMENT: What prior knowledge is needed for this task?

Identify the prerequisite skills and understandings.

- Define virtue: honesty; mutual respect; attentiveness.
- Determine understanding of the elements of construction: blueprint, materials, budget, final product.
- Assess understanding of the engineering design process and determine if review is required.
- Analyze understanding of budget development and cost worksheet.

FORMATIVE: How will I measure student progress toward understanding?

R

RIGOR: How can I increase the student's cognitive thinking?

Identify tasks that can elevate student thinking, improve inquiry, and increase conceptual understanding.

- Students determine the sequence of design steps to create the blueprint and the art pedestal structure.
- Allow for possible blueprint revision once construction begins if design is not plausible.
- Students determine the constraints for the structure. Must be no more than eighteen inches tall, no wider than a foot.
- Students make list of possible materials for construction, determine a price for each one, and establish a budget based on estimate of materials needed.

E

EXCITE: What is the hook to excite the learner?

Create the scenario to engage the learner.

The town has just received word that a very wealthy citizen is going to donate a beautiful piece of art to the community. The town will need to have a pedestal built to support the art piece. The Board of Selectmen for the town decided to hold a competition for community members for the design and construction of this pedestal.

ENGAGE: How will the students be cognitively engaged throughout the unit?

List the STEM practices that will be used as evidence.

- Students will ask questions to define the plan for the blueprint.
- Students will construct a viable argument for defending the group's blueprint design and how it will meet the

SECONDARY STANDARDS (Passengers)

Engineering/Design

ETS1.A: Defining and delimiting engineering problems
Possible solutions to a problem are limited by available materials and resources (constraints). The success of a designed solution is determined by considering the desired features of a solution (criteria). Different proposals for solutions can be compared on the basis of how well each one meets the specified criteria for success or how well each takes the constraints into account.

Mathematics

7.EE.B.3: Solve multistep real-life and mathematical problems posed with positive and negative rational numbers in any form (whole numbers, fractions, and decimals), using tools strategically. Apply properties of operations to calculate with numbers in any form; convert between forms as appropriate; and assess the reasonableness of answers using mental computation and estimation strategies.
7.G.B.6: Solve real-world and mathematical problems involving area, volume, and surface area of two- and three-dimensional objects composed of triangles, quadrilaterals, polygons, cubes, and right prisms.

- How can the engineering design process be used to find a solution to the problem?
- How will first drawing a blueprint of the design help with the construction of the pedestal?
- Is there more than one design that will provide for the best solution to the problem?

Classes 2–3: Assign working groups—explain their first task is to develop a blueprint agreed upon by the team for their group's piece of art. Pass out the bags containing the small stuffed animals. Show the materials they will be using and explain the budget sheet. Students then develop their group's blueprint, which will need to be approved by the head of the board of directors (the teacher) prior to beginning construction.

Class 4: Once blueprint is approved, provide the students with the art pedestal performance rubric and have them begin construction.

Class 5: Student groups develop their presentation they will give to the board of directors.

Classes 6–7: Students give presentations and team rubric assessment review.

Establish the assessment tools you will use to monitor progress and inform instruction.

- Provide opportunities for individual and group check to monitor collaboration based on team assessment rubric.
- Check in with individual students on their ability to apply scale design to the development of the completed pedestal.
- Students are able to explain in their own words the development process of the group's art pedestal and its connection to the engineering design cycle.

SUMMATIVE: What criteria are needed for students to demonstrate understanding of the standards, content, and skills?

Create a checklist of criteria for use in a rubric.

The student teams will give a group presentation explaining their process for developing their art pedestal, which includes the blueprint, budget, and steps to construction.
Criteria for evaluation:

- performance rubric
- team assessment rubric
- budget summary sheet

- Students decide on the presentation format to highlight the features of their completed art pedestal.

RELEVANCE: Does the learning experience provide for relevant and real-world experiences?

Identify current topics and local issues that can make the tasks more engaging.

- Have local historian address student questions about the following using pictures of statues in the community: Why does the pedestal look like it does? Why was the statue erected? Why was it important to the community at the time? What were the objections to the erection of these statues?
- Students develop and construct an authentic blueprint, create a budget, and build a structure, applying their skills and knowledge across disciplines and working together within a team.

approval of the Town Board of Directors.

- Students will apply mathematics and the engineering design process to construct their art pedestal.
- Students will communicate, based on evidence of a sound structure, why their group's design is the best.

EXPLORE: What activities will help students address the Driving Question?

List questions for students to investigate that will lead them to a deeper understanding of the content and skills.

- How does our community govern itself and make decisions? (Take a tour of city hall.)
- What types of pedestal shapes support different statues and large pieces of art? (Look at famous statues from around the world.)
- How are they constructed?
- How will developing an accurate, to-scale blueprint help with the construction of their group's piece of art (graph paper, rulers, calculators)?
- What materials will best suit the construction for your group's pedestal, which will be cost-effective and also attractive?
- Can your group explain the application of the engineering design process for the construction of the art pedestal?

References

Achieve. 2015. *Closing the Achievement Gap: 2014 Annual Report on the Alignment of State K–12 Policies and Practices with the Demand of College and Career.* Washington, DC: Achieve.

Achieve Inc. 2013. "Next Generation Science Standards. For States. By States." www.nextgenscience.org.

Banilower, E., K. Cohen, J. Pasley, and I. Weiss. 2010. "Effective Science Instruction: What Does Research Tell Us?" http://www.centeroninstruction.org

Barkos, L., V. Lujan, and C. Strang. 2012. *STEM—Catalyzing Change Amid the Confusion*. Portsmouth, NH: Center of Instruction.

Bell, P., B. Lewenstein, A. W. Shouse, and M. A. Feder, eds. 2009. *Learning Science in Informal Environments: People, Places, and Pursuits.* Washington, DC: National Academies Press.

Carroll, Lewis. 1946. *Alice in Wonderland*. Kingsport, TN: Kingpost.

Covey, S. 1989. *Seven Habits of Highly Effective People*: New York: Simon & Schuster.

Dierking, L. D. F., and J. H. Falk. 1994. "Family Behavior and Learning in Informal Science Settings: A Review of the Research." *Science Education* 78 (1): 57–72.

Froschauer, L., ed. 2016. *Bringing STEM to the Elementary Classroom*. Arlington, VA: NSTA Press.

Harland, Darci J. 2011. *STEM Student Research Handbook*. Arlington, VA. NSTA Press.

Hord, Shirley M., William L. Rutherford, Leslie Huling-Austin, and Gene E. Hall. 1987. *Taking Charge of Change*. Washington, DC: Association for Supervision and Curriculum Development.

Innovative Technology Experiences for Teachers and Students. April 2013. *Program Director's Guide to Evaluating STEM Education Programs: Lessons Learned from Local, State, and National Initiatives.* Arlington, VA: National Science Foundation.

International Center for Leadership in Education. 2003. www.leadered.com.

Jackson, R. 2011. *How to Plan Rigorous Instruction (Mastering the Principles of Great Teaching)*. Alexandria, VA: Association for Supervision and Curriculum Development (ASCD).

Keeley, P., F. Eberle, L. Farrin, and L. Olliver. 2005. *Uncovering Students' Ideas in Science—Volume 1*. Arlington, VA: NSTA Press.

Loucks-Horsley, S., K. E. Stiles, S. Mundry, N. Love, and P. W. Hewson. 2010. *Designing Professional Development for Teachers of Science and Mathematics*. Thousand Oaks, CA: Corwin.

Marks, Helen. 2000. *Student Engagement in Instructional Activity: Patterns in the Elementary, Middle, and High School Years*. American Educational Research Journal. Spring, Vol. 37, No. 1, pp. 153–184.

McTighe, J., and D. Reese. 2013. *Understanding by Design & Defined STEM*. http://images.definedstem.com/docs/UBD-Defined-STEM.pdf.

National Council for the Social Studies (NCSS). 2013. *The College, Career, and Civic Life (C3) Framework for Social Studies State Standards: Guidance for Enhancing the Rigor of K–12 Civics, Economics, Geography, and History*. Silver Spring, MD: NCSS.

National Math + Science Initiative. 2013. *Whirligig Lollapalooza—Exploring Science and Engineering Practices*. www.nms.org NMSI: Dallas, TX.

National Research Council. 2000. *How People Learn: Brain, Mind, Experience, and School: Expanded Edition*. Washington, DC: The National Academies Press.

———. 2011. *Successful K–12 STEM Education: Identifying Effective Approaches in Science, Technology, Engineering, and Mathematics*. Washington, DC: The National Academies Press.

———. 2012a. *Education for Life and Work: Developing Transferable Knowledge and Skills in the 21st Century*, edited by J. W. Pelligrino and M. L. Hilton. Washington, DC: The National Academies Press.

———. 2012b. *A Framework for K–12 Science Education: Practices, Crosscutting Concepts, and Core Ideas*. Washington, DC: The National Academies Press.

———. 2013. *Monitoring Progress Toward Successful K–12 STEM Education*. Washington, DC: The National Academies Press.

———. 2014a. *Developing Assessments for the Next Generation Science Standards*, edited by J. W. Pellegrino, M. R. Wilson, J. A. Koenig, and A. S. Beatty. Washington, DC: The National Academies Press.

———. 2014b. *STEM Integration in K–12 Education Status, Prospects, and Agenda for Research*, edited by M. Honey, G. Pearson, and H. Schweingruber. Washington, DC: National Research Council.

National Science Board. 2007. *A National Action Plan for Addressing the Critical.* Arlington, VA: National Science Foundation.

Newmann, F. M., G. G. Wehlage, and S. D. Lamborn. 1992. *The Significance and Sources of Student Engagement.* Student Engagement and Achievement in American Secondary Schools, edited by F. M. Newmann, pp. 11–30. New York: Teachers College Press.

New York Post. *35 of Yogi Berra's Most Memorable Quotes.* September 23, 2015. http://nypost.com/2015/09/23/35-of-yogi-berras-most-memorable-quotes/

SRI Education. 2015. *Measuring the Monitoring Progress K–12 STEM Education Indicators: A Road Map*, edited by B. Means, J. Mislevy, T. Smith, V. Peters, and S. Nixon Gerard. Arlington, VA: SRI Education.

Stein, Mary Kay, Margaret Schwan Smith, Marjorie A. Henningsen, and Edward A. Silver. 2009. *Implementing Standards-Based Mathematics Instruction: A Casebook for Professional Development, 2nd ed.* New York: Teachers College Press.

STEM 101: Intro to Tomorrow's Jobs. Occupational Outlook Quarterly. Spring 2014. www.bls.gov.

Vasquez, J. A., C. Sneider, and M. Comer. 2013. *STEM Lesson Essentials, Grade 3–8: Integrating Science, Technology, Engineering, and Mathematics.* Portsmouth, NH: Heinemann Press.

Wallace, J., Sheffield, R., Rénnie, L., Venville, G. 2007. Looking Back, Looking Forward: Re-Searching the Conditions for Curriculum Integration in the Middle Years of Schooling. *The Australian Educational Researcher*, Volume 34, Number 2, August 2007, 29–40.

Wiggins, G., and J. McTighe. 2005. *Understanding by Design (Expanded 2nd Edition).* Alexandria, VA: ASCD (Association for Supervision and Curriculum Development).

Wolf, K., and E. Stevens. 2007. "The Role of Rubrics in Advancing and Assessing Student Learning." *The Journal of Effective Teaching.* University of Colorado at Denver and Health Sciences Center. www.uncw.edu/cte/et/articles/Vol7_1/Wolf .pdf.